THE OPTICAL MUNITIONS INDUSTRY IN
GREAT BRITAIN, 1888–1923

T0330526

Studies in Business History

THE OPTICAL MUNITIONS INDUSTRY IN GREAT BRITAIN, 1888–1923

BY

Stephen C. Sambrook

Routledge
Taylor & Francis Group

LONDON AND NEW YORK

First published 2013 by Pickering & Chatto (Publishers) Limited

Published 2016 by Routledge
2 Park Square, Milton Park, Abingdon, Oxfordshire OX14 4RN
711 Third Avenue, New York, NY 10017

First issued in paperback 2015

Routledge is an imprint of the Taylor & Francis Group, an informa business

© Taylor & Francis 2013
© Stephen C. Sambrook 2013

BRITISH LIBRARY CATALOGUING IN PUBLICATION DATA

Sambrook, Stephen C.
The optical munitions industry in Great Britain, 1888–1923. – (Studies in business history)
1. Optical industry – Military aspects – Great Britain – History – 19th century.
2. Optical industry – Military aspects – Great Britain – History – 20th century.
3. Optical instruments – Great Britain – Design and construction – History – 19th century. 4. Optical instruments – Great Britain – Design and construction – History – 20th century.
I. Title II. Series
338.7'6814'0941-dc23

ISBN-13: 978-1-138-66196-7 (pbk)
ISBN-13: 978-1-8489-3312-5 (hbk)

Typset by Pickering & Chatto (Publishers) Limited

CONTENTS

PREFACE AND ACKNOWLEDGEMENTS

This book has its somewhat unlikely origins in my longstanding, but for some time dormant, interest in the seemingly disparate technologies of weapons and optical instruments. That interest was reawakened by chance several years ago when a friend gave me a copy of Anthony Pollen's *The Great Gunnery Scandal* which he thought might make for a good book at bedtime. Telling the story of one inventor's efforts before the First World War to sell the British Admiralty a new system for controlling long-range naval gunnery, it did indeed turn out to be what used to be called a 'jolly good read'. I owe Michael Grey a good deal of thanks for the gift of the book because it led me to the realization that the two technologies not only converged in the control and deployment of weapons but also had distinct commonalities in the way they were conceptualized, understood, selected and acquired by their military clients. As I began to explore both areas it quickly became apparent that although there was an abundance of literature, both scholarly and popular, on the design and manufacture of weapons, there was virtually nothing on the production of optics for warfare.

Optical munitions have indeed been very much been overlooked by historians of all persuasions, something which was pressed home to me by my then undergraduate supervisor Alan Simmonds. It was he who first encouraged me to investigate the subject and to develop my findings. Without his urging, I might never have progressed much further down the road, but the journey which I then started led to my meeting Ray Stokes at the University of Glasgow's Centre for Business History in Scotland who shrewdly sought to convince me that the industry which made optical munitions was even more interesting than its artefacts and a far more fruitful area for research. He was, of course, absolutely correct and I acknowledge cheerfully that his vision of military optics was then better focused than mine. Without his counsel and guidance this book would not have taken the shape that it did. I owe an equally great debt of gratitude to my wife, Miriam, whose enthusiasm and support have never wavered and who showed the patience of a saint in tolerating my frequent distraction and sustained inattention; without her the task would have been immeasurably harder.

I must also thank and pay tribute to the staff at the libraries and archives, which provided so much source material. The University of Glasgow's Archive Service must head the list, if only because I spent so much time there that I got to know both the institution and its staff better than any other. I am especially indebted to the University Archivist for permission to quote extensively from the Barr & Stroud collection and for the use of its photographic illustrations.

Other archives which provided the essential source material include the National Archives at Kew, London; Bromley Library's Local History Section; Cambridge University Library; Cooke Optics Ltd; Cumbria County Archives; Hampshire County Record Office; Leeds Industrial Museum; Leicester County Record Office; the Ministry of Defence Admiralty Library; the Science Museum Library and the University of York's Borthwick Institute for Archives. All these have my gratitude but some individuals call for especial mention.

Alison Brech at the Borthwick Institute for Archives spent time and energy way beyond the call of duty cheerfully and enthusiastically following up ideas for me. Barbara Lowrie of the ZGC Corporation very generously gave me access to the private archive of Taylor, Taylor & Hobson Ltd and provided yet more information from her own collection of material relating to H. Dennis Taylor. She also introduced me to the design staff at Cooke Optics Ltd who explained the bewildering complexities of manual optical computation and showed me the tools used for doing it. William Reid was more than generous with his time and hospitality on numerous occasions, not just providing insights into the British Army's early use of optical devices and about the working of the military mind but also showing me examples of optical munitions from his own extensive collection of instruments. His kindness was particularly appreciated.

I have received a great deal of advice, guidance and support in the preparation of this story and any errors which exist in its telling remain entirely my own responsibility.

LIST OF FIGURES AND TABLES

LIST OF TECHNICAL TERMS

Binocular: vision with both eyes, producing the effect of depth; an optical instrument using two telescopes which are fixed together, each for one eye, in order to present the operator with a sense of three-dimensional vision.

Depression rangefinder: a distance measuring system used by coastal artillery. The instrument was located at a known height above sea level and aimed at the waterline of a target vessel in order to measure the angle of depression. A mechanical calculator then automatically produced the range to the target.

Dial sight or panoramic sight: a sophisticated sighting system for artillery incorporating three types of prism and a telescope in the form of a periscope.

Galilean binocular: also known as a 'field glass'. A hand-held binocular of simple non-prismatic construction. It was much less expensive than a prism binocular and usually of low magnification.

Mekometer: a type of rangefinder operated by two observers who were separated by a known distance, with each taking readings from a common target in order to calculate its distance.

Monocular: pertaining to one eye; an optical instrument designed to be used with one eye.

Optical glass: types of glass specially formulated and made for incorporation into lenses or prisms for optical instruments. Such glasses need to be entirely free from physical blemishes and inherent stress and come in many different types to allow the design of optical systems free from aberrations, which tend to degrade image quality.

Periscope: a device used to displace the line of vision, usually in a vertical direction. Normally associated with submarines, where it is a highly complex device incorporating two telescope systems in tandem combined with prismatic elements. It allows above-surface observation from a submerged vessel. It was also used in simpler form for observation from trenches during the First World War (1914–18).

Prism: a transparent optical glass body with polished plane surfaces inclined in relation to each other. Made from optical glass, prisms are used to direct, to shorten or off-set light paths within an instrument. They are essential in periscopes and rangefinders where complex forms are frequently required. Their most common use is in prismatic binoculars.

Prismatic (prism) binoculars: an observation instrument using two joined telescopes incorporating a prism system to shorten the instrument's length and increase the separation of the telescope's entry lenses to enhance the effect of three-dimensional vision.

Rangefinder: an optical instrument to measure distance between itself and a selected object. There are two basic types. In the *coincident type*, two separated telescope systems provide an optical base whose images are combined in a single eyepiece. The operator typically sees an image divided horizontally with the upper and lower parts displaced relative to each other. Manipulating a control aligns the two and a range reading is thus obtained from a scale. Coincident rangefinders use triangulation to perform this task. In the *stereoscopic type*, the two telescopes' images are combined in a binocular system in which the operator sees just one image with a range mark which appears to 'float', moving towards or away from the operator in response to the rotation of a control. The range is obtained by manipulating the control until the floating mark seems to 'sit' on the selected target. The potential accuracy of both types of instrument is proportional to the length of its optical base and the magnifying power of its optical system. The British rangefinders mentioned in this volume are all of the coincident type.

Sighting telescope: a magnifying device adapted to be used in the aiming of ordnance (for example field guns). It incorporates an aiming mark known as a graticule or reticle and is attached to the gun in such a way as to be insulated from the shock of firing.

Telescope: an optical instrument which makes distant objects more clearly visible by magnification using an optical system.

Telescopic sight or riflescope: a telescope adapted specifically to the aiming of a rifle used by an individual. It incorporates an aiming mark to enable a precise sighting of the weapon and, normally, some means of adjustment for the range to target. The telescopic sight has to be able to withstand the recoil of the rifle on which it is mounted.

INTRODUCTION

This book describes and analyses the creation and growth of a successful technological and strategically vital manufacturing community in Britain whose story runs counter to perceptions of the general relative decline in British scientific and technological industries during the later nineteenth and early twentieth centuries. The optical munitions industry has remained largely unnoticed by historians, not least because of assumptions that its products were essentially little different from those of the general optical manufacturing sector. In fact, the optical munitions industry produced a highly specialized, increasingly sophisticated and complex group of products in response to other advancing technologies that not only influenced military and naval weapons design, but also the strategy and tactics of their use. These devices were specially designed or deliberately adapted for use in warfare and were used for a multiplicity of tasks, from observation and sighting individual weapons to controlling gunnery on land and at sea. They ranged from derivatives of the simple terrestrial telescope to highly complex and sophisticated apparatus such as the naval rangefinder and the submarine periscope, creations without which the principal strategic weapons of the First World War could not have functioned at all. This relatively small but highly specialized manufacturing sector attained a crucial degree of importance within the armaments industries so that its story adds substantially to the understanding of the performance of specialized technological manufacturing in Britain between the late nineteenth century and the mid-1920s.

Despite its importance, this industry has been generally overlooked, misunderstood or wrongly located within British industrial manufacturing. There are a number of interconnected reasons for this. First, before 1914 the industry was relatively small despite its increasing importance to the armed forces. Only a few businesses were active in the field and just one of them operated on a truly substantial scale. The second reason is that the nature of its products is hardly likely to catch the historian's eye. The massive ordnance on a battleship is hard to miss, as are the field guns and the rifles of any early twentieth-century army; both the ships and the soldiers are an immediate prompt that a substantial industry must exist to support them. But optical munitions are largely invisible. The optical devices which measure the range settings for the battleship's guns often seem to be part of the ship's structure, the optical sights for artillery are lost in the complexity of the gun's mounting and the binoculars used to reconnoitre the distant enemy are hidden amongst the soldiers' accoutrements. The artefacts are out of sight and the industry which makes them, seldom, if ever, brought to mind. The third, and in many ways the most important, reason for the lack of recognition is that the identity of the optical munitions industry has been effectively camouflaged by the very term 'optical'. Can optical instruments for warfare really be so different to those used in science and medicine, or even for leisure activities such as photography, that they demand recognition of a distinctly separate manufacturing industry? The answer is resoundingly 'yes', as this book shows.

Having said that, the small number of historians of business and technology who have touched on the subject have tended to assume that military optical devices were indeed little more than modifications of instruments intended for civil use, so fostering the notion that optical munitions constituted a relatively small part of the output of the scientific instruments community and were essentially no more than an adjunct to it. Some of the reasons for the conflation of civil and military optical manufacturing can be laid firmly at the door of the first printed account dealing with munitions manufacture generally.

The *History of the Ministry of Munitions* was prepared in the early 1920s as an official record of its work during the First World War and includes a section covering optical manufacturing.[1] Never intended to be an analytical exercise, it focused on the achievements of the Ministry without delving deeply into the background of the industries it dealt with. The official account allocated just forty-four pages to the subject in a twelve-volume work, covering what it described as 'The Optical and Scientific Instrument Trade'. Two principal problems have arisen from this and helped to misdirect the few scholars who later touched on it. First, it located optical munitions production firmly in the commercial scientific instruments industry, a position which has subsequently been accepted without question. And second, it maintained that the production of optical munitions before mid-1915 took place within the context of a backward and inefficient optical industry, an impression that has also been generally adopted as conforming to later 'declinist' views of British technological industries. This book takes issue with both those positions.

There have been only a few instances of historians coming across the production of optical munitions and giving it their attention. One of those is Roy and Kay MacLeod's 1977 book chapter which looks at it in the context of the British government's relationship with the optical industries generally during the First World War.[2] They certainly saw optical munitions production as part of the general optical instruments industry's activities principally in the context of the state's mediations in what they described as the 'science-based industries', considering that such production took place 'at the extreme end' of that sector.[3] Following the lead given by the Ministry's account, they emphasized the necessity of state aid during the war to overcome the inadequacies endemic in the declining pre-war optical industry.

In another rare reference to optical munitions production, Mari Williams compared the British and French 'precision industries' between 1870 and 1939, in the context of connections between precision engineering and the military sciences.[4] Although not dealing specifically with optical manufacturing, she recognized the potentially catalysing links between such industries and the armed forces,[5] citing the Glasgow rangefinder maker, Barr & Stroud, as an instance of how such connections could shape the growth of a business.[6] Nevertheless, the company was again seen as part of the instrument making community rather than being closer to the armaments industry.

A third reference to optical munitions manufacture is in Anita McConnell's history of the York optical instrument maker, Thomas Cooke Ltd.[7] She clearly showed the firm's involvement in optical munitions production before the First World War, but because much of that was either adaptations of, or derivations from, earlier types of survey instruments designed for the civil market, she regarded them as essentially the same as the firm's other commercial optical apparatus.[8] She also saw optical munitions manufacture as an adjunct to commercial instrument production rather than a separate enterprise.

Historians of military technologies and warfare have also overlooked the importance of optical munitions, especially in the naval context where the significance was greatest and most clearly to be seen. There, optical devices have typically been seen as subordinate elements of other elaborate weapons systems, rather than as critically important artefacts in their own right. This is surprising, given that even before 1914 front-line warships depended entirely on optical devices to identify friend from foe, to measure distances for setting sights and for laying the guns on their distant targets. The oversight is still more pronounced in the case of the submarine, which depended wholly on an extremely complex optical device for its warlike capabilities.

Jon Sumida's study of the introduction of gunnery direction systems into the Royal Navy before 1914 emphasized the mechanical computational elements that predicted the future position of a moving target, minimizing the role played by optical instruments in providing the initial target distance required to set the process in train.[9] Although acknowledging the presence of the rangefinder, he passed over the significance of optical devices in systems of gunnery control, with scant attention to the essential need for telescopic sighting and other observation devices in them.[10]

The lack of attention to optical munitions has been still more marked with the submarine. There, the periscope was the only means by which an underwater vessel could see what was happening above the surface and thus function as an effective weapon. Norman Friedman's examination of the evolution of the US submarine is a rare recent instance of the role of the periscope in the vessel's evolution being mentioned, but his discussion concentrated on its installation and maintenance in the boat's hull.[11] Like Sumida, he relegated optical instrumentation to a secondary role within a larger technological system.

This sustained lack of recognition by historians to appreciate fully the growing significance of optics in both the tactics and strategy of warfare before 1914 and during the First World War has led to a failure to investigate this specialized sector of the precision industries. As this book shows, the optical munitions field was really closer to the armaments sector rather than that of civil scientific instruments manufacturing. Making and selling optical munitions took place in an environment quite unlike commercial manufacturing where 'onerous conditions and hyper-critical inspection' by the clients was a constant pressure on the makers.[12] Investigating the industry reveals that, like much of the arms industry, it was far removed from the picture of relative decline that has been painted for the British optical industries in general. In a nutshell, the variegated make-up of the optical sector has so far escaped recognition.

The optical munitions industry's story has much to command the attention of those interested in business history, in military and naval history, and the history of technology as well. It depicts a specialized sector of British industry that even before 1914 was technologically advanced, by no means lacking in scientifically trained staff and was progressively, even aggressively, managed. It was also able to meet all the needs of Britain's armed forces and to sell successfully overseas against its continental rivals. Much of this challenges existing perceptions of inadequacy not only in the optical munitions sector but also in the more general optical instruments sector as well. This account explains the flawed base on which those beliefs have largely been built. The industry's history also illustrates how important it was to the state, directing attention towards the policies and motivations of the Admiralty and the War Office in the way they selected and equipped themselves with optical instrumentation before the First World War and how they regarded the makers on whom they came to rely.

The highly specialized products of the optical munitions industry had, with very few exceptions, no scope for outlet into any civil marketplace. Its only clients were the governments of states which, increasingly in the decade before the start of the First World War, sought to update or improve the equipment of their armed forces in response to the rapidly evolving technological progress in military science. In consequence, competition for business was generally governed by factors additional to the normal commercial bases of technical superiority or competitive pricing. Besides the need to master the demanding technologies involved in designing and manufacturing optical munitions, selling these products demanded its own peculiar set of skills. Dealing with armies and navies – and hence their political masters – called for a combination of salesmanship, technological sophistication and diplomatically inspired negotiation including, increasingly, an understanding of international affairs. In Britain the industry's relationship and interaction with the state played a crucial role in the way it become a strategically vital sector of British manufacturing industry even before the start of the First World War.

The story of the industry's evolution embraces not only elements of entrepreneurship and invention, of business history and the growth of business structures – the internal elements – but also the evolution of military technologies and the connected aspects of domestic and international politics – the external elements. Although the principal theme is of an evolving industry which became a strategically critical part of what would now be called the defence infrastructure, there are other themes from the external elements which add layers to the story, adding to its complexity but providing access to a more extensive and reliable understanding. These include the exploration of how particular groups within military and naval societies influenced the choice of individual technologies; the state's attitude to the rights of inventors; the international proliferation of armaments; the factors influencing expenditure on armaments at both macro and micro levels and, not least, the willingness of the British government to sustain vital elements of the armaments industries in times of peace and financial stringency.

Telling the Story

The starting point of 1888 has been chosen not because optical munitions sprang into being then, but because it marks the first occasion on which the British state publicly sought the involvement of the civil sector in seeking the design of a particular optical artefact for military use. The end date of 1923 similarly does not signify that optical instruments for warfare suddenly vanished from the earth. Rather, it signifies the point at which the political will to engage in disarmament and economic stringencies effectively brought to an end the widespread and regular series manufacture of optical munitions in Britain. That pushed the surviving elements of a greatly contracted industry into a hiatus which amounted to a state of hibernation where it remained until the re-armament programmes of the later 1930s. By that time, electronic technologies were starting to emerge that would eventually make obsolete some of the previously most important parts of the industry.

The industry's history between 1888 and 1923 divides itself neatly into three segments. The first covers the inception and maturing of the industry up to the start of the First World War in 1914. It explains why optical munitions manufacture became necessary to the armed forces and considers how and why the British state clients – the Admiralty and the War Office – evolved such substantially different attitudes to the industry. This section also shows how the optical munitions sector became distinctly different to civil and scientific instrument production as it expanded in response to growing demand at home and abroad.

The second segment looks at the industry's performance in war between 1914 and 1918, showing how it responded to the massive expansion of demand and the organizational disorder in the first months of the conflict. It then goes on to analyse the effect of the Ministry of Munitions' involvement and both illustrates and explains the agenda for changing the entire optical industry which has subsequently coloured perceptions of it. Case studies of particularly important artefacts are used to illustrate and assess the effectiveness of wartime production.

The final part examines what happened to the industry after the war's end. It takes the story through the upheaval of industrial demobilization and shows how the consequences of peace impacted on the expanded wartime industry. The final chapter of this segment looks more deeply at how the key players in the industry coped with the evaporation of business in the climate of economic retrenchment and disarmament.

Sources

Putting together this story would have been impossible without access to the contemporary records of the players who take part in it. Archive material for much of the optical industry is frustratingly scarce; the records of some of the important firms have vanished completely, leaving behind little more than fragmentary collections of trade catalogues and advertisements. Business records for firms which were well established even before 1888 and which stayed in business for almost a century after have disappeared, presumably consigned to bonfires by their liquidators and the demolition crews which cleared the sites of their factories. Such is certainly the case for many of the firms which feature in the story, and accurately piecing together their roles is consequently not easy. But, fortunately, enough company records to make the task possible have survived, and some of those amount to 'treasure troves'. Two of those especially deserve mention.

Much material has been preserved for Thomas Cooke & Sons Ltd of York and now resides in the Vickers Instruments Archive at the University of York's Borthwick Institute for Archives. Cooke's were by no means the largest of the optical munitions makers but they were engaged in advanced optical design work from the 1890s. In the first decade of the new century they became involved in the development of a highly sophisticated naval gunnery control system which incorporated a complex and novel type of rangefinder. The firm's drawing office records provide a chronology of its evolving optical munitions designs and its carefully classified collection of British and foreign patents shows how assiduously it monitored developments in the field. That the company ran its military and civil work within the same factory by no means amounted to a conflation of purpose; the conception of the work was clearly divided both in the boardroom and on the shop floor.

Still more important than Cooke's reference material is the archive of Barr & Stroud, the largest of the optical munitions makers. Its material is unrivalled in its scale and scope. Dating back to 1888, the firm kept records of all its outgoing correspondence – and much more besides – from the early 1890s until it was eventually taken over in the 1980s. Happily, the successor companies not only kept the mass of material but passed it over for safe keeping to the University of Glasgow where it now resides. The huge collection of letters sent out from the early 1890s provides almost an *embarras de richesses*, providing both a macro and a micro view of what the firm did and how it went about it. Equally informative are the order records, detailing not only every order received by the firm after 1901, but also the progress of the job even through to the receipt of payment. Given the scale on which the business operated and the wealth of data, it is not always easy to prevent the story being skewed towards the one company. There is, fortunately, other material available elsewhere which acts as a balance.

The disequilibrium is countered to a large degree by the existence of official Admiralty and War Office records which for the period up to mid-1915 provide a picture not only of what was being bought but also the motivation for its selection. For the period up to 1919, the surviving and still extensive material left behind by the Ministry of Munitions provides information beyond what might be expected in company records, especially concerning the armed forces' contemporary perceptions of the optical industry as a whole. When the Ministry was closed down after the war, its records were heavily 'weeded out' and in some cases the surviving files are incomplete, creating frustrating gaps. This is particularly the case with contract records which generally seem to have been discarded, but to redress that imbalance many of the weekly technical reports detailing dealings with individual companies were retained. It is sifting through the minutiae of those that uncovers the complexity of managing a wartime industry and which counters the later official account portraying the Ministry as the omniscient saviour of a backward and ailing industry. After the war, the Admiralty and War Office records again flesh out the state's attitude towards the industry in the years up to 1923.

Even with the mass of archive material available, though, there are still some things which cannot be seen clearly, if at all. That is more particularly true for much of the pre-war British optical instrument industry generally. Given the disappearance of so many of its company records it is difficult to know the scale on which it actually operated or the extent to which scientific training actually figured in its activities.[13] It is somewhat ironic that the optical munitions sector should shed more light on the general instrument making sector than vice versa.

Technical Terms and Measurements

Although this is not a history of technological artefacts, repeated mention of different sorts of optical munitions is unavoidable and some of the terms used will possibly be unfamiliar to readers. The amount of 'jargon' has been minimized but it is impossible to eliminate it entirely. It may therefore be helpful to refer to the short section dealing with technical terms before proceeding further than this introduction.

There are some instances where past prices are accompanied by modern values to give a better impression of the monetary scale of business being done. Such conversions can be done in a number of ways but for the sake of simplicity, and because they are used only as an illustrative guide, the figures given have all been obtained from one source as cited in the references, and on the same basis.

Distances and dimensions do appear regularly though, and they are given in the same units found in the original sources. These are usually imperial units and in the context of this book it is sufficient to treat one inch as twenty-five millimetres, one foot as thirty centimetres, and yards and metres as the same. Any discrepancies will be of no practical concern here, although those who used optical munitions in earnest would be right to demur about the imprecision.

1 THE EMERGENCE OF THE INDUSTRY, 1888–99

The story of the emergence of the optical munitions industry from 1888 to 1899 is largely about the growing importance of one instrument – the rangefinder – and the influence which the state had on the emergence of an industry for the manufacture of such specialized optical devices for use in war. The state's influence was transmitted through the activities of the War Office and the Admiralty, both of whom showed a common commitment to the idea of using optical aids but differed significantly in how they organized their acquisition and deployment. These differences changed over time and were based on a number of sometimes complex issues which included technological and tactical considerations, along with other, social, factors – reflecting particular aspects of what were really two very different military societies. That assortment of disparities meant that they proceeded along very different lines in taking up optical munitions and in the way they related to the industry on which they would increasingly come to rely. The War Office, although the first mover in taking steps that might stimulate the growth of a new industry, proved to be less inclined to seek innovation and often reluctant to move forward in the adoption of new types of optical devices.

Even if the term optical munitions would have been unfamiliar to the armed forces in the 1880s, optical instruments intended to give soldiers and sailors some form of tactical advantage in warfare had been used regularly on a small scale since the seventeenth century. The military significance of the telescope as a means of identifying objects too far distant to be seen clearly by the unaided eye was recognized as early as 1609 and by the late nineteenth century the terrestrial telescope had been generally recognized as an indispensible observation instrument on land and at sea. However, despite this recognition of the potential of optics to make the prosecution of warfare more effective, there were still very few other items of optical apparatus in general use with the British armed forces. The reasons for this were essentially straightforward and rooted in contemporary military technologies and tactics. The effective ranges of small arms and artillery were still sufficiently short that targets could easily be seen by the unaided eye and ranges usually estimated with enough accuracy to deliver fire effectively. Rudimentary magnifying sighting telescopes for rifles and artillery had been produced after the Crimean War, as well as some rangefinding devices for field guns and coastal artillery, but these were used only in small numbers or experimentally and were generally derived from instruments sold in the civil market place, chiefly for land surveying.[1] In 1888 neither the British Army nor the Royal Navy were supplied with any standardized or systematically produced patterns of optical apparatus specifically produced for military or naval use. Optical munitions still did not exist as a recognized category of 'warlike stores' in the armed forces, nor was there an industry geared to the regular manufacture of optical instruments for use by them.

Optical apparatus was used only on a relatively small scale. Soldiers and sailors employed telescopes for identifying and observing a distant enemy and reading flag or semaphore signals, but they were issued in limited numbers and were little different to those sold commercially. Telescopic gun sights were rarely used by either service and rudimentary rangefinders were found only in the Army. This began to change after the late 1880s, not because of any radical progress in optical science, but because developments in armaments technologies created problems in maximizing the potential of new weapons, obstacles which could be overcome by the application of existing optical technologies. As the range and accuracy of guns grew, it became increasingly essential to know target distances more precisely in order to set elevations correctly, and to have some means to aim at targets so far distant as to be almost invisible to the eye. These needs were actually not novelties in the 1880s, having first appeared more than thirty years previously. The gradual introduction of more accurate small arms and artillery had prompted both services to experiment with rangefinding instruments and aiming telescopes in a haphazard manner since the 1860s, but only with the arrival of the new more powerful 'nitro' or smokeless propellants did the conditions finally emerge where optical aids to gunnery became not just desirable but essential. As a result, the War Office and the Admiralty began separately to seek new instruments, so creating conditions which could nurture the growth of a distinctive optical munitions industry. From the standpoints of design and manufacturing, by far the most challenging of this first generation of gunnery instruments was the rangefinder, and this chapter examines its evolution in the light of the influences affecting both its design and production. Innovation in armament technology came to drive specialized optical manufacturing into a completely new sector that sat between civil optical instrument production and the armaments industry.

The War Office and the Rangefinder

To the War Office must go the credit for being the first mover in the process of bringing into being an optical munitions industry, although, as will become clear as the story progresses, it did little to galvanize that industry into expansion and profit in the following two decades. In May 1888, anticipating the introduction of a new, more powerful magazine-fed infantry rifle that offered the possibility of being used effectively at ranges even beyond 1,000 yards, the War Office published an advertisement in the London journals *Engineering* and the *Engineer* inviting designs for a rangefinder suitable for use by the Infantry.[2] The Army already used small numbers of rudimentary range-measuring devices modelled on instruments and techniques that were well established in surveying and map-making.[3] These were mostly found in the artillery and used the principle of triangulation, setting out a baseline of known length perpendicular to the object whose range was required, and then measuring the angle subtended between the target and the extremity of the base in order to calculate the target's distance. All were essentially modifications of the usual surveyor's tools which had been devised by serving officers as private projects and adopted only on a small scale. Their small-scale manufacture had been done by instrument makers within their normal business of production for the civil market so that military contracting had never been more than a peripheral activity within the optical instrument making community.

The advent of the new rifle presented the possibility of infantry units delivering high volumes of fire at ranges so long that, without knowing the distance correctly, errors in sighting would mean missing even massed ranks of men.[4] If the artillery units were indifferently equipped with rangefinders in 1888, then the infantry forces were even worse off. Their 'Field Rangefinder', which was based on the surveyor's box sextant and had been designed by an artillery officer, had proven far from satisfactory and was criticized even within the Army for its inconvenience, fragility and inaccuracy.[5] These shortcomings, if irksome, were of relatively little consequence when combat distances for rifle-fire were usually short enough for sight settings to be less than critical, but the prospect of engagements at greatly increased range meant that a more reliable means of assessing distance was now essential. The decision was therefore taken to seek a new rangefinder, and a public request was made for submissions in May 1888. The public advertisement was made because the War Office lacked both an infrastructure for the design or production of optical devices and, unlike other areas of armaments production, had no existing close relationship with the relevant industry.

What the Army's infantry units wanted was clearly spelled out in the advertisement. The successful design had to deliver a rangefinder hardy enough for use on active service in all weathers, to be portable by one fully equipped soldier and require no more than two men to operate it. It also had to be able to measure ranges with an accuracy of 4 per cent at 1,000 yards, the largest permissible error to ensure appropriate sight settings.[6] Designs had to be received before 1 August 1888, which allowed just eight weeks from the date of publication for the preparation of submissions.

There was a substantial British optical instruments industry that was well capable of considering the problem. At least thirty-four manufacturers existed in 1888 who were making, or had recently made, survey instruments, telescopes or microscopes, all of which employed precisely worked optical systems variously incorporating lenses, prisms and mirrors mounted in protective housings.[7] These firms made up an industry producing a wide range of precision-made artefacts employing both optical and mechanical engineering skills which sold not only domestically and throughout the Empire, but also in Europe, the Far East, the United States of America and South America. Its products included telescopes, from small hand-held opera and field glasses to complete astronomical observatories, surveying instruments from the simple box sextant up to the largest transits for primary surveys, laboratory microscope bodies and their eyepieces and objectives, as well as stereoscopes, spherometers, ophthalmoscopes and lenses for photographic cameras and lantern-slide projectors. A panoply of contemporary optical instruments was being made in Britain by a demonstrably diverse and capable industry which might reasonably have been expected to produce viable ideas for a new military rangefinder.

That did not happen. Despite this substantial infrastructure, not one firm entered a design of its own in the trials held in 1889, and the only civilian submission came from private inventors.[8] Given the range and level of expertise regularly demonstrated by British optical companies, this signal lack of response by the industry seems at first difficult to understand. Its explanation can be found in the combination of the way the War Office published its requirement and how it conducted its business.

The invitation for designs was published as a small advertisement in journals aimed at the mechanical engineering community, which was perhaps not the ideal place for the War Office to seek an optical device. If noticed at all by the instrument makers, it allowed them only eight weeks for the submission of designs, a period which gave prospective entrants relatively little time to design a novel artefact and prepare drawings, let alone to construct a prototype in order to evaluate its viability. There was no indication of the scale of business which might result from any successful design and, as will become apparent as the story progresses, the War Office could be a difficult client to deal with when it came to matters of reward. The question of payment even for patented designs lay entirely in the hands of the state. The Patents, Designs and Trade Marks Act of 1883 allowed any government department to use a patented design without any previous settlement of terms for payment. If the owner of the intellectual property in the design objected to what was offered, the only course of redress was to the Treasury, another department of state, with no provision for further appeal. Any business which was acquainted with the War Office's terms and conditions of business might have felt that they amounted to a disincentive rather than an inducement for competition and innovation. It is not too surprising that the optical industry's response was underwhelming. Two principal contending designs eventually emerged, one from a serving artillery officer who was already a prolific designer of rangefinding devices, and the other from two academics who had neither prior expertise in designing measuring instruments, nor in dealing with the War Office.

Archibald Barr (1855–1931) and William Stroud (1860–1938) may have seemed unlikely participants in the competition, but they would nevertheless go on to become key figures in the growth and success of optical munitions manufacturing in Britain; they, or rather their business, will be met repeatedly in following chapters. In 1888, they were employed by the Yorkshire College in Leeds; Barr was Professor of Engineering and William Stroud was Cavendish Professor of Physics.[9] Neither had any connection with the armed forces or the optical industry, and Stroud's later description of their decision to enter the competition suggests it was rooted in a combination of academic frustration and momentary caprice rather than any serious desire to become military contractors. The two men had met in Leeds during August 1885 and they subsequently formed a friendship which resulted in the design of a camera to simplify the production of lantern-slides as teaching aids. This had some small commercial success and early in 1888 they decided to collaborate again, this time in a research project intended to enhance their academic standing. They had made very little progress by the time the War Office advertisement appeared on 25 May and Stroud subsequently recorded that

> On the morning of May 26th, Dr Barr came round to see and proposed to drop the subject of the determination of the mechanical equivalent of heat and take up the intervention for Rangefinders, about which neither of us knew anything. In blissful ignorance of what had already been done on the subject, we dashed off regardlessly.[10]

Stroud's account, written late in his life and meant to be read by his family as an informal memoir, suggests a levity not mirrored in their subsequent actions. Whatever optimistic enthusiasm they began with was quickly tempered by the 'ghastly failure' of their hastily constructed rudimentary prototype, after which they begin to approach the problem in a more deliberate and scientific manner.

Both were convinced from the outset that what would best meet the conditions of the War Office's advertisement was a self-contained device able to be used by one man. Commissioning a search for patents on rangefinding devices led to finding a number of earlier, relatively unsuccessful designs. In 1860, the Scottish instrument maker Patrick Adie had been granted British Patent 37/1860 for a device incorporating 'improvements in means to measure angles', and a second one in 1863 (British Patent 608/1863) for developments of his first design. Adie's rangefinder had been included in trials held by the Royal

Artillery in 1869, although it had performed poorly and was dismissed as fundamentally unsuited for army service, principally on the grounds that its design differed radically from what artillerymen were already seeing as an optimal pattern.[11] The Adie design was a self-contained type operated by one man, with a short measuring-base of 3 ft 6 in. (approximately 1 m) rather than the 75 ft or more used with those based on surveying practice.[12] The inferior accuracy of the Adie rangefinder was ascribed by those testing it to the inherent limitations of its short measuring-base rather than to technical shortcomings in its construction. Two other designs of short-base rangefinders had been patented in the mid-1880s, one by H. R. A. Mallock (British Patent 8043/1885) and the other by the Astronomer Royal, W. H. M. Christie (British Patent 12404/1886). Both incorporated refinements of Adie's earlier model but neither had been marketed.

Looking at the patents convinced the two professors that the principal reason for the failure of those earlier rangefinders had not been their relatively short measuring-base, but rather the weakness of the mechanical engineering around one key part in the optical system. Previous attempts had all provided a distance reading by aligning a movable image of the target with one that was fixed to the operator's view. The displacement of the movable image had been done by rotating either a lens or a mirror about its vertical axis, but the amount of rotational movement was so small that reading errors resulted through what have been colourfully described as 'drunken screws' and 'deranged reflectors'.[13] Errors of a few thousands of an inch in the positioning of such components resulted in reading errors so large as to nullify the instrument's purpose. Stroud's solution to that problem involved the replacement of one type of optical component with another, coupled with a new mechanical arrangement to house it. Instead of rotating a lens or a mirror, the image displacement could be done by moving a wedge-shaped prism along the optical system's axis, converting a rotational movement of a few thousandths of an inch into a much longer lateral motion through a system of gearing which could be arranged to provide a direct reading of the target's distance (see Figure 1.1 on p. 15).

Having identified a way forward, the substantial commercial potential of their evolving design, as well as possible problems in its exploitation, seemingly became very quickly apparent to Barr who promptly queried the War Office about what would happen to the intellectual property in the design if it were taken up by the Army.[14] Displaying a lack of knowledge of how the War Office went about its business, he asked whether it would become government property, or if the inventor would 'be at liberty to treat with foreign governments'. Perhaps naively, Barr also enquired if it would be 'necessary or advisable' to seek patent protection for the invention. The dismissive reply gives some additional insight as to why established companies may have been disinclined to respond to the War Office's request for rangefinder designs. Barr was told that arrange-

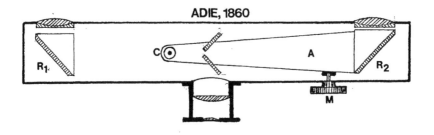

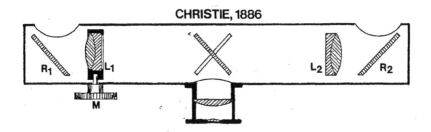

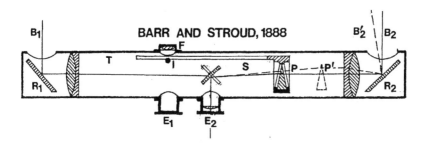

A. Arm

B. Beam of Light

C. Pivot

E. Eye-piece

F. Finder

I. Index

L. Lens

M. Micrometer Screw

P. Prism

R. Reflector

S. Ivory Scale

Figure 1.1: The Barr and Stroud Rangefinder design of 1888 with two earlier types by Christie (1886) and Adie (1860). Illustration courtesy of University of Glasgow Archive Service, GB0248/UGD295.

ments about ownership and rights would be made only after the trials, without any indication as to how the questions of title and reward would be dealt with. The inference was that the War Office regarded itself as the sole arbiter of how inventors should be treated and would offer only the terms it thought fit. Furthermore, it refused even to consider any claims for expenses in developing prototypes until they had been submitted for trials, and gave no guarantee that costs would be met if the submission was unsuccessful.[15] As for patents, Barr was told, inventors should decide for themselves about the desirability of protection. The professors quickly made up their minds about that issue and promptly lodged a provisional specification at the Patent Office to cover what they termed the 'tracking prism' arrangement. Neither the use of a prism nor moving an optical component along its horizontal axis were actually novel but their application and combination in a distance measuring device were, and the 'invention' could therefore be patented and protected. Undeterred by the War Office's somewhat peremptory reply, and having secured protection from anyone (other than a government department) subsequently seeking to usurp their idea, the two then submitted their design to the War Office which was sufficiently impressed to request a sample for testing in trials the following year.[16]

Barr's early instinct about the rangefinder's potential saleability seems to have developed rapidly as he and Stroud worked to develop a prototype for trials. Ideas for improving the design came quickly as they progressed up their learning curve, and Barr began lobbying to gain access to the Trials Committee to persuade them to accept an instrument built to a new design. By December 1888 he had somehow obtained letters of introduction to the senior reviewing officer which led to a meeting on 18 January 1889.[17] He and Stroud had by then completely redesigned the optics and their housing to produce a more easily used and durable instrument which they wished to build and substitute for the one recently delivered.[18] The Committee refused this because it contravened the terms of the competition, but Barr was told that if the original design passed the first tests, their improved version might possibly be entered in later trials.[19] The original prototype was tested in March and performed well, enabling Barr to persuade a new senior reviewing officer to follow the lead of his predecessor and allow the submission of detailed proposals for the modified design. In mid-June the Ordnance Committee which was to conduct the testing agreed that the new model could be entered for the second set of trials in August, provided it was delivered by the end of July.[20]

The professors' new design employed considerable sophistication in design to enhance the instrument's performance. Stroud's original design used pentagonal prisms for its 'end-reflectors' in place of the mirrors in earlier rangefinders which he considered a fundamental weakness. Prisms, which had only started to become common in optical instruments during the 1880s, were more stable and resistant to the distortions that were almost impossible to avoid in glass mirrors and had repeatedly affected the accuracy of earlier rangefinders. Stroud's proposed optical improvements used a novel 'objective prism' that had one face ground to a curve to let it also function as a lens as well as a reflector. Replacing the earlier separate object lens and pentagonal prism assemblies with an integrated unit not only improved the image's contrast and brightness but also provided an assembly that could be more rigidly mounted, thus substantially improving both the optical and mechanical performance.[21] However, the novel concept was unfamiliar to optical manufacturers and even more difficult to make than a plane-surfaced pentagonal prism, problems which led to serious consequences in the course of testing at Woolwich Arsenal.

Sometime after the March trials, it seems the professors began to fear that their new rangefinder's high selling price might count heavily against it.[22] Their chief competitor had emerged as a design by a serving officer, Major H. S. Watkin, which was not greatly dissimilar to other earlier efforts originating with the Army and was, again, modelled on civilian surveying practice. His two-operator, long-base 'Mekometer' was less convenient and slower to use than Barr and Stroud's rangefinder but, because of its relatively simple design and construction, could be supplied at a price which was scarcely one-sixth of the professors' instrument. According to Stroud's memoir written in the 1930s, the fear of being eliminated on cost grounds prompted them to look for substantial savings in production costs which eventually led to the submission of a version which not only abandoned the sophisticated objective prisms but also reverted to the use of silvered mirrors as end-reflectors.[23] This summary is difficult to reconcile with the terms of the competition and almost certainly presents only a partial picture of the reasons for the shift back to components which were already known to be less than satisfactory.

Barr and Stroud had justified the submission of a new instrument to the trials committee on the grounds of the significant optical and mechanical advantages obtained through using objective prisms, so justification for the change from them to glass mirrors is not easy to understand. The problems with mirrors were well known to Barr and Stroud, who had been quick to recognize that their use had been partly to blame for the failures of earlier rangefinders. The relatively thin sheets of glass were virtually impossible to mount in a manner that let them respond to temperature changes without flexing and distorting, which resulted in false range readings. Even if the economics had been great enough to bring the price down to anything like the Mekometer's, the dangers of using them still remained. But reducing the rangefinder's cost so greatly was highly improbable, because even without any optics the complexity of its strain-resisting body inevitably meant a price greater than the simple sheet-metal box used by Watkin in conjunction with an optical system that was little more than two mirrors and three lenses similar to those used in spectacles. Despite Stroud's later account, the change could not have been driven solely by cost considerations.

A likely explanation involves the abilities of the British optical instruments industry generally and specifically the complex prisms that were at the heart of the improved model. The two professors were entirely reliant on outside sources for the manufacture of their optical components because their workplace, the Yorkshire College, had no facilities for making either lenses or prisms. The trials rangefinders' optical systems incorporated spherical lenses similar to those already being used in high quality terrestrial telescopes and survey instruments, which could be ground to order by any company familiar with making precision optical elements. The pentagonal prisms used as end reflectors in both designs were not so easily obtained because few instruments yet had optical systems employing them and high precision 'flat grinding' was a speciality employed by only a small number of makers. The well-established York business of Thomas Cooke & Sons reputedly made up both types of components for the first trial rangefinder,[24] but it is unlikely that Cooke's were tasked with the unfamiliar objective prisms for the second model. A more likely source was the London firm of Adam Hilger & Co.

By 1888 it had established itself as the country's leading maker of prisms, and Archibald Barr would already have been acquainted with its owners, Adam and Otto Hilger, through his earlier connections with Lord Kelvin in Glasgow.[25] The Hilgers had made optical parts for Sir Archibald Campbell, an associate of Kelvin, since 1875, and in 1888 Otto Hilger had moved to Glasgow to work for Campbell in his laboratory. Barr had been Kelvin's assistant before he went to Leeds, and maintained contact with him during the time he and Stroud were developing the rangefinder, sending him details of the design and receiving comments on it.[26] The notion of the Hilger company having been asked to make the prisms for the 1889 instrument is supported by a reference in the firm's surviving papers that although the first order from the partnership of Barr & Stroud came in 1891, there had been other earlier ones individually from the two professors.[27] Hilger's prism expertise was considerable, but it was a small firm which, as will be seen in subsequent chapters, regularly had difficulty keeping its work on schedule, and in 1889 relied almost entirely on Adam Hilger himself for the most difficult and exacting work such as Stroud's objective prisms. If their unusual form had caused problems and delayed the rangefinder's completion, then an impending delivery deadline of 31 July would have demanded an urgent solution for which the use of mirrors was perhaps the only course open. Whatever the reason, the behaviour of the rangefinder at its trial was indubitably a catastrophe. When Stroud used it in the cool of the August morning it worked well, but as the day progressed the sun's heat had the predictable effect on its mirrors, causing it to produce readings which were very inaccurate.[28] After the trials, Barr and Stroud were probably unsurprised to be told that their instrument had been rejected and that the competitor they had seen as their major obstacle, Major

Watkin's Mekometer, had been chosen for adoption. However, it is by no means certain that Barr and Stroud's rangefinder would have won the competition even if it had performed more accurately in the hot August sun.

Even though the Army still lacked a satisfactory distance-measuring instrument in 1889, what amounted to a rangefinding paradigm, or philosophy, had gradually become entrenched in the Army's thinking since the 1860s. One part of this mindset was that only a derivation of the surveyor's two-observer long-base system of measurement could produce a sufficiently high standard of accuracy to be acceptable. This was based on trials which showed that when used deliberately under test conditions they did indeed produce demonstrably more accurate readings than the self-contained single-observer designs.[29] In a lecture at the Royal Artillery Institute in 1881, reviewing every variation of rangefinder tried since 1861, the speaker emphasized that trials had repeatedly established the unsuitability of 'yard telemeters', by which he meant the self-contained, single-observer types with a measuring-base of only 3 ft (90 cm).[30] He emphasized that what was needed was refinement and simplification of the long-base type and insisted that no other system was likely to perform satisfactorily. The other part of the mindset was that experienced soldiers were the most likely source of suitable patterns for service use, a sentiment apparently borne out by every model yet used having been designed by an officer in the Royal Artillery. Watkin's Mekometer not only fitted into the notion of an improved long-base type, but also came from someone whose credentials as a designer of rangefinders could hardly have been bettered from the War Office's point of view.

Watkin had an impressive record as an inventor of distance measuring devices. In the 1870s, when still a captain in the Royal Artillery, he had designed a rangefinder using the principles of surveying specifically for defensive coastal artillery at a time when increasing emphasis was being laid on the use of fortresses to repel invading forces.[31] Its selection process had been protracted, with trials taking place between 1876 and June 1881 when it was finally adopted and put into service 'under a cloak of great secrecy'.[32] The Watkin Depression Rangefinder was a sophisticated derivation of the surveyor's level, intended to be used in gun batteries which were sited well above sea level. It could only be used from a permanent mounting whose height above a mean sea level was precisely known and which formed the extended measuring-base for the system. The name came not from any psychological effects it had on its operators, but because it measured the angle of depression between it and the target in order to complete the automatic calculation of the target's range. Unlike any other rangefinder then in use, it proved both reliable and consistently accurate, enhancing both its designer's reputation and its principle of operation.

Because the depression rangefinder was a fixed unit it was both easy and quick to operate, but the infantry's Mekometer was by no means so simple to use.[33] Its two operators needed to stand erect and apart at an exactly pre-determined distance – usually 25 yds – on level ground in order to establish the necessary horizontal base length, and then had to align their separate instruments on a pre-determined target. The principal operator then manipulated his instrument vis-à-vis the other until images reflected from the target and the other instrument were brought into alignment in the principal's unit, enabling a range reading to be subsequently worked out. Unlike the single-observer type, where all that was needed was to aim the instrument and align two images by means of an operating wheel, or the Depression Rangefinder, where a sighting telescope was depressed into alignment with the target, Mekometer operators needed coordination and perseverance, characteristics that were later to create serious problems in action and greatly limit its utility.

Having chosen its new infantry rangefinder by September 1889 in a clear endorsement of the value of optical instruments in battle, the War Office then went on to display a marked reluctance to commit itself to actually putting the device into service. It took no immediate steps towards ordering it, which did nothing to help either the infantry or benefit its chosen manufacturers, Thomas Cooke & Sons of York.

The War Office and Thomas Cooke & Sons

The Mekometer was not officially introduced into service until October 1891, although Cooke's were told much earlier that they were to be the makers.[34] The decision to give Cooke's the job of manufacturing it displays another facet of the War Office's approach at the time to procuring optical munitions. The choice was made not because the company had won any competitive tendering process but because it had an existing relationship with Watkin and the War Office that went back to the 1870s in connection with the Depression Rangefinder. As would later be the case with many other optical munitions, mechanical engineering skills were as important as optical ones in making the instrument. Its optical component was a high magnification telescope incorporating an aiming mark that was similar to those which Cooke's already produced for survey instruments; the device's accuracy and dependability came from a complex system of precisely machined cams and gears that automatically translated the depression angle into a range reading and target bearing for the guns. Watkin had approached the company to build his early prototypes as much on the basis of Cooke's reputation for mechanical engineering as for their optical prowess, and the relationship had been cemented as successive versions were trialled by the War Office. The reputations of designer and maker alike had benefitted substantially from the device becoming a key component in the network of coastal forts and batteries that had been set up since the 1860s as a strategic defence system against attack and invasion, where it was vital to the doctrine of engaging an enemy at long range before he could approach closely enough to deploy his own armament.[35] Watkin's device was considered so important to the state that he was awarded the very substantial sum of £25,000 in 1888, for 'transferring his patents to the War Office'.[36] Whilst Watkin may have benefitted handsomely from the Depression Rangefinder, neither this earlier optical munitions product nor the new Mekometer would bring substantial financial reward to Thomas Cooke & Sons.

Although Cooke's had a substantial reputation for the quality of its products and both a large domestic and export trade for survey instruments and telescopes, it also had a longstanding record of liquidity problems. They had begun in the late 1860s and were still pressing in the early 1890s, chiefly caused by losses from the mismanagement of large contracts for astronomical telescopes.[37] By 1890, despite a generally expanding market for survey apparatus, profits from general instrument making had become inadequate to cover losses on the astronomy side and the firm was in financial difficulties, seriously exceeding the limit on its bank overdraft. In those straitened circumstances, Cooke's came to see optical munitions manufacture as a potential solution to its financial difficulties.[38]

In September 1890, Cooke's current account with the Yorkshire Banking Co. was overdrawn by almost double its agreed limit of £1,000.[39] By then the firm had been told by the War Office that it would, at some still unspecified date, receive a contract to supply the Mekometer. On the strength of this, Cooke's asked for its overdraft to be raised to £6,000 to cover current liabilities, and for a further £4,000 to purchase additional premises. Cooke's told the bank that the money was required until February 1892 when 'the account would be put in credit by moneys to come to them from the Government', presumably from existing orders for the Depression Rangefinder and new ones for the Mekometer. The bank turned down the application for such a massive increase in borrowing, although it continued to tolerate Cooke's persistently large overdraft on the strength of the assurance that substantial funds would eventually result from government contracting. However, instead of being in credit by February 1892 as promised, the firm continued to be overdrawn and in June the firm was forced to ask for its borrowings to be increased for three months to £5,000. The reasons for this are not shown in the bank's or the firm's surviving records, but it is most likely that either the War Office business had not matched expectations or its other trading had deteriorated markedly. Whatever had been the company's expectations of government contracts, the optical munitions work had not been able significantly to improve the company's overall financial condition.

Assuming that the firm was being honest with its bankers, the company was either being excessively optimistic about its prospects with the Mekometer, or misled or let down by the War Office concerning the likely scale of business. The firm was told in September 1890 that it was going to be given orders and had prepared the necessary drawings to commence manufacture by April 1891.[40] Production, though, went ahead very slowly, with the School of Musketry commenting in its Annual Report for 1892–3 that the only rangefinders in service with the infantry were still the older types which the Mekometer was intended to supersede.[41] The implication is that large orders had not been placed, but it also begs the question of how many Mekometers Cooke's had actually expected to supply. That is not easy to determine as no relevant company documentation has survived. However, the War Office's contemporary scales of issue for infantry rangefinders indicate that only 300 Mekometers were likely to have been ordered, plus a small number for spares, so that with a selling price of £7, its total value could scarcely have exceeded £2,500.[42] For the instrument to generate enough profit to clear the requested £10,000 of borrowings by early 1892, contracts for around 4,000 would have been required.[43] If Cooke's had genuinely expected orders of that level, the disappointment must have been considerable, because the Mekometer was never actually ordered in bulk and percolated into service so slowly that even by the start of the Second Boer War in 1899 neither the infantry nor the artillery were fully equipped with it.[44]

Although the War Office made an early move towards adopting a rangefinder, it did relatively little to encourage the development of an optical munitions industry during either the 1880s or the 1890s. Besides being slow to put the Mekometer into general service the Army showed a marked hesitancy to adopt, or even evaluate, any of the radically improved observation devices which began to be available by the mid-1890s. In 1893 the German firm of Carl Zeiss at Jena launched a then novel family of telescopes employing complex prismatic optical systems which for the first time provided high magnification with wide fields of view and enhanced stereoscopic vision. These properties were combined in robust, compact housings which readily lent themselves to military use. The new prismatic binocular telescopes were not officially evaluated, despite the substantial advantages they offered and their early endorsement and adoption by the German army.[45] Although many of the Zeiss instruments' features were patented, some key aspects were not, and broadly similar ones could have been manufactured in Britain had the War Office chosen to take any of them into service.

Far more stimulus for creating optical munitions manufacture came from the Admiralty in this period, particularly through its efforts to find a rangefinder that began shortly before the British Army decided to adopt the Mekometer.

The Admiralty and the Rangefinder

Unlike the Army, the Royal Navy had never officially employed any instruments for measuring distances at sea and consequently had evolved no philosophy on rangefinding. The Navy was eventually directed towards the subject by the same emerging weapons and propellant technologies that had influenced the War Office, as well as the growing attention to naval policy after 1884 that led to the large spending programme of the Naval Defence Act of 1889.[46] Faced with the prospect of new ships armed with improved guns which could shoot further and quicker, the Admiralty became concerned with maximizing those benefits and decided to investigate whether a satisfactory rangefinder for shipboard use could be obtained.[47]

In 1889, the Royal Navy still depended on the War Office for its supplies of guns and their related equipment, and so it was to the Director of Artillery at Woolwich that the Navy first took the issue of rangefinding.[48] The first Infantry rangefinder trials had just been held and the second set was due to start soon when the Navy's Ordnance Committee announced in June that 'the question of a naval rangefinder is one of pressing importance'. The Navy's gunnery school, HMS *Excellent*, was also asked for an expert opinion. Its captain's reply said he attached little or no importance to rangefinders because a better way of finding ranges rapidly was by observing the splashes made by misses from the new quick-firing guns that were then being introduced. In view of that, he thought there was no need for the Ordnance Committee to pursue the matter. The Director of Naval Ordnance, however, was little impressed with the gunnery school's con-tinued espousal of what amounted to the tactics of the early nineteenth century and on 9 July told the Director of Artillery that it would be a 'great advantage' to have an effective ship-borne rangefinder and asked for the benefit of his expe-rience. The Director passed on the names of the entrants to the current trials but offered the Admiralty neither advice nor comment.[49] A year later, the Naval Ordnance department formally asked Woolwich's Director to set in motion the process of finding a suitable naval rangefinder, and in early April 1891 sent a set of conditions which the successful instrument was expected to achieve.

Although the War Office was responsible for instigating the selection process, it took no part in the actual trials which were organized and conducted entirely by the Navy, and there were significant differences in the way that the Admi-ralty competition was managed. Rather than simply placing advertisements and waiting for a response, the Admiralty identified likely inventors, including the entrants to the 1889 War Office trials and specifically invited them to submit designs.[50] The technical demands were considerably greater those for the Army trials: the device had to measure ranges to an error no greater than 3 per cent at 3,000 yds, irrespective of ship motion, speed or the course of either its own vessel or its target; it had to take ten readings per minute (to produce a mean range); needed provision for 'some system of instantaneous communication' to send the data to the guns, and it had to be as simple and as durable as possible to withstand conditions at sea. What the Admiralty was asking for was not just a rangefinding device but the nucleus of a complete system for gunnery control.

Submissions were made by private inventors and serving officers and, as in the previous War Office trials, the two strongest contenders again emerged as Barr and Stroud and Major Watkin. His submission was a derivation of his successful depression rangefinder. This incorporated an electrical circuit which supplied two angle-readings from widely separated points to a central control station which, like the depression type, converted them into ranges for transmission to the guns.[51] This attempt to transfer the Army's rangefinding to naval use proved unsuccessful and was officially rejected due to its inferior accuracy when tested at sea, despite the lengths it went to providing an integrated solution for gunnery direction. The fact that it demanded the installation of a whole system on board ship may also have counted against it because of the need for installing extensive electrical cabling to a control room, both of which were likely to be expensive and difficult to arrange with existing ships.[52] The Barr and Stroud design, on the other hand, was far less complex and in the form submitted made little impact on a ship's structure. It was only 5 ft long and could be used as a 'stand alone' item capable of being moved around from one mounting point to another, almost like a telescope.[53] Unlike Watkin's there was as yet no data transmission system, only the promise of delivering one once the Admiralty's detailed requirements about shipboard locations for rangefinders had been settled. Despite that, the professors' newly improved model so outperformed its competitors in trials in April and June 1892 that it emerged as the clear winner.

Barr and Stroud – from Cottage Industry to Manufactory

Winning the competition was still no guarantee of the rangefinder's adoption or financial reward. On 10 June 1892 the Admiralty wrote to Barr about what might happen next about the rangefinder.[54] The letter raised the question of terms and conditions in the light of the inventors' proposal to sub-contract manufacture and asked what 'you and Mr Stroud [*sic*] are willing to accept for these instruments' should the Royal Navy adopt them. There were three points which the Admiralty wanted answered before it would consider whatever price the inventors might ask. Was the rangefinder patented? Had its details been made public? Could secrecy be guaranteed? The final issue would materially affect the terms to be offered. If secrecy could be assured, then the government would be interested in acquiring the sole rights to the rangefinder, either by a lump sum or a royalty on each one bought. Because the most important aspects of the rangefinder's design had been patented and so were already in the public domain, all notions of secrecy immediately fell by the wayside, but after some delay the Admiralty nevertheless asked the inventors to offer their patents to the Crown, presumably in an attempt to prevent proliferation amongst foreign powers. Barr was reluctant to relinquish control of the rangefinder and his subsequent demand for £75,000 in view of their 'enormous commercial potential' was so high as to amount to a diplomatic refusal.[55] As the Admiralty's main preoccupation was secrecy rather than buying up a potential commercial investment, there would have been little chance of Barr being paid such a large sum, irrespective of the accuracy of his valuation. Having set aside any ideas of buying sole rights, at the end of November the Admiralty asked Barr to quote for six rangefinders for further trials marking the start of a commercial relationship that was to outlast the naval life of the optical rangefinder. The professors were now on the road from inventors to successful entrepreneurs.

The road to entrepreneurial success was not without its pitfalls and problems. In February 1893 Archibald Barr and William Stroud had neither the means to make the instruments' parts themselves nor even a workshop in which to assemble bought-in components.[56] Barr was by then Regius Professor of Engineering at the University of Glasgow and Stroud still at the Yorkshire College in Leeds, a separation of over 200 miles which did nothing to facilitate the production of the first batch of rangefinders for the Royal Navy.[57] The instruments to be made reflected the research done by the pair since 1888, which had, in the meantime, produced another five rangefinder-related patents.[58] The new model had advanced considerably since 1889, having a longer measuring-base of 5 ft, a stronger double-tube body of non-ferrous metals to avoid influencing ships' compasses, and finally a revised optical system that avoided the use of the expensive and still hard-to-obtain large end-reflecting prisms by substituting speculum-metal reflectors that were intrinsically more stable than glass mirrors. The components were all made by outside contractors and gathered together for final assembly in Glasgow, this time not on university premises but in Barr's own home under circumstances which amounted to little more than a cottage industry.

The fabrication of mechanical parts was done by James White & Co. in Glasgow, but the optical work was spread between Adam Hilger in London and Chadburn Brothers in Sheffield, Yorkshire.[59] Chadburn's was an important supplier of the simpler optical components in Barr and Stroud's rangefinders from at least 1892 until well into the First World War. Founded late in the eighteenth century, the business made a range of optical instruments, as well as large quantities of lenses for the ophthalmic trade.[60] Although the main telescope part of the rangefinder demanded high-grade lenses, its aiming viewfinder and some other parts could be made satisfactorily with the simple lenses used in spectacles, and it made no sense to pay Hilger & Co. for higher quality components when they were not necessary. For the more sophisticated achromatic lenses and the complex eye-piece prism assemblies, orders continued to go to Hilger & Co. who supplied some of the components directly to Glasgow and others to Stroud in Leeds.[61] He then built up the complicated central arrangement of prisms and lenses that presented the separate images to the operator's eye before despatching each finished component to Glasgow for assembly in the rangefinder body and for final adjustment.

This method of sourcing and assembling components was adequate only with orders for small numbers of instruments and Barr was aware that as demand grew a significant transformation of organization and methods would be required for the business. The first stage in the progression from cottage industry to a manufacturing organization came when Barr persuaded one of his own university students, Harold Jackson, to abandon his studies and become his full-time, salaried administrative and technical assistant in 1893.[62] Although then only twenty-one, Jackson quickly came to occupy a key role in the progress of the business and to play a part scarcely less important than Barr himself. He will be encountered repeatedly in succeeding chapters. The second step was the negotiation of a sales agency agreement in the same year with the Newcastle-on-Tyne armaments maker and warship builder, W. G. Armstrong Mitchell & Co.

For an informal partnership of two academics whose total business to date amounted to just one order for six rangefinders, the need to set up an international marketing structure so soon may seem premature. However, as William Stroud noted later, once the Admiralty had announced its decision to buy rangefinders, foreign interest rapidly burgeoned and 'within a few months enquiries poured in from places as far apart as Tokio [sic] and Washington'.[63] Naval and military attachés arrived in Glasgow to see the instrument and the prospects for foreign business quickly seemed encouraging, not just to the inventors but to outsiders in the armaments business as well. In late March 1893 Barr was approached by Armstrong's who asked whether he was actually able to supply rangefinders and, if so, at what price.[64] Having quoted a figure, Barr followed up by enquiring in May whether or not Armstrong's would be able to get any orders. That seems to have prompted the firm's Commander E. W. Lloyd to arrange a meeting at which he put the question of an agency to Barr.[65]

Lloyd, who had a considerable reputation as an artillery expert, had recently retired from the Royal Navy and already knew about the Barr & Stroud rangefinder, having mentioned it in his recently published work *Artillery: Its Progress and Present Position*. When writing the book he was still – like the captain of HMS *Excellent* some two years earlier – unconvinced of the instrument's value.[66] However, when the Admiralty's decision triggered foreign interest, Armstrong's recognized that orders were likely from their overseas clients and moved to concentrate the export sales of the rangefinder in their own hands rather than any competitor's. The discussions with Lloyd led Barr to prepare a draft agreement which he returned to Armstrong's after some amendments in early July with an accompanying letter that said:

> One of my chief concerns for desiring to come to some such agreement as you had proposed was that we had not the machinery for securing prompt payment in the case of business being done with the Smaller States. This, Captain [*sic*] Lloyd said would be no difficulty to you and I understood that you are willing to undertake the financing of foreign business in so far as the securing of payment is concerned.[67]

Barr may have been stimulated by the idea of foreign sales, but financing them would indeed have been problematic. By 1893 the partners' expenditure on research and patenting had become considerable, and according to Stroud they 'were now approaching the end of our financial tether'.[68] Their losses so far were at least £1,247, all of which except for £300 had come from their own resources.[69] Barr's income at Glasgow University in 1893 was £468,[70] and Stroud's salary at Leeds was unlikely to have been greater, so that their situation in the absence of outside financing must indeed have been difficult. Work done for the British government was relatively safe in the sense that payment was guaranteed and might perhaps be financed through eking out suppliers' credit terms, but the risks of foreign sales were a different matter, as Barr's letter to Armstrong's made clear.

The ten-year agreement that was signed in September 1893 was potentially advantageous for Barr and Stroud's future development.[71] Armstrong's would promote their rangefinder to the exclusion of any other by influencing the foreign powers who were their clients for ships or guns. Barr and Stroud would fix the selling prices, and Armstrong's would guarantee payment within three months of taking delivery, irrespective of whether or not the foreign client had paid for them. In return, Armstrong's would earn a 12.5 per cent commission on all sales from foreign enquiries except – at Barr's insistence – those from Germany.[72] Almost at a stroke, and with no outlay, the embryonic business had acquired both a foreign sales department and a guarantee of payment within a set time limit, a combination to be well pleased with.

Despite the expectation that Armstrong's would generate new trade, few orders resulted until well into 1894, when Barr and Stroud's total business amounted to just thirteen rangefinders, eight of which were for the British Admiralty. Despite Stroud's insistence that there had been much foreign interest, there was as yet little material benefit from it. Navies, including the Royal Navy, were still to be convinced that the rangefinder worked efficiently or was even necessary, and Barr was keen to get the Admiralty to commit itself in order to gain what he rightly saw as a valuable endorsement.[73] In late May, Jackson asked the Admiralty whether the Navy thought the rangefinder was 'suitable' for adoption, and encouraged a favourable answer by saying further improvements had been made and offering a discount of 10 per cent for orders of fifty or more.[74] Eventually, in February 1895, the Admiralty confirmed that the instrument had been 'definitely adopted' for use by the Navy, although there was no indication of the extent to which it might be ordered.[75]

Foreign interest was still not stimulated to the point where significant orders were being placed and the stance of the Imperial German Navy perhaps sums up contemporary attitudes. In April 1894, the German Naval Attaché in London, Captain Tülick, had asked about delivery of a sample to Berlin, asking if someone could be sent to demonstrate it 'without charging anything, or only a moderate sum'.[76] The response to that has not survived but a fortnight later, undeterred by whatever he had been told, the Attaché wrote again to enquire 'if you are doing any other interesting work for the British Admiralty'. Irrespective of whatever he learned from that attempt at espionage by post, a rangefinder was subsequently ordered, to be collected from Glasgow by a German 'expert' in July, but only after payment had been made. The correspondence illustrates some of the problems that Barr was having promoting sales. Tülick had asked to visit what he thought was a factory in Glasgow to see for himself not only the rangefinder but whatever else was being made there, and he also wanted a firm delivery date for the one just ordered.[77] The reply, sent over Barr's signature but from its style composed by the youthful Jackson, neatly juggled assurance and embarrassment. First, Barr and Stroud were specialists; they only made rangefinders, a subject with which 'few men are acquainted'. On delivery times, the 'peculiar nature' of the work meant that 'unforeseen accidents might slightly retard completion', but once finished it would be best to gather as many experts in Berlin as could be managed at one time to show them the rangefinder. As for a factory visit, the letter confessed 'We have not a workshop of our own, except a small one in Professor Barr's house' and that 'the important parts' were made in various outside locations. This did not deter German interest, and the rangefinder was duly delivered on time and sent to Berlin where, instead of Barr being able to show it off, it was (according to Stroud) immediately 'forwarded to Zeiss to be copied'.[78]

Orders were slow to come in from sources other than the Admiralty and income remained modest through to the end of 1895. By then just fifteen range-finders had been ordered by foreign clients, compared to thirty-three by the Royal Navy, adding up to a total value of £6,125.[79] This slow growth was influenced by both the Admiralty and by the professors still being in the early phases of developing both its application and design. For the Navy, the question was one of how to employ the rangefinder, and for Barr and Stroud the problem was how to refine the instrument to produce a satisfactory product that could be marketed with the endorsement of large-scale adoption by the British Admiralty.

Up to 1895 each instrument delivered differed slightly from its predecessor.[80] Only in that year was the 'FA2' model introduced, representing the arrival at a developmental plateau where a standardized product could be manufactured to a fixed specification rather than individual examples being modified as they were produced.[81] In April 1895 the Admiralty asked for a quotation for twenty,[82] so that with the design having reached a stage of stability and a substantial order from the Admiralty, it finally became feasible to advance the development of the business by acquiring, for the first time, workshop premises and operatives to do part of the manufacturing. The previous year, the inventors had created the formal partnership of 'Barr & Stroud's Patents' to exploit the value of the designs already registered.[83] That had allowed for either licensing or manufacturing, but by early 1895 the partners' attention was concentrated on the latter, not least because the earlier question of producing some means of transmitting data electrically from the rangefinder to the ship's guns had been resurrected by the Admiralty.

The original specification had called for the eventual provision of such a system even though it had not been required at the trials, and in November 1893 the Admiralty had finally asked for the submission of the necessary 'electrical apparatus'.[84] A set of these Range and Order (R&O) instruments was manufactured and tested by mid-April 1894, but no decision about them had been made when, in early 1895, Armstrong Mitchell had raised the question in the context of incorporating them into ships under construction.[85] The possibility of extra business coming from Armstrong's sooner than from the Admiralty must have impressed on Barr the increasing urgency of having some properly organized workshop premises of his own.

Although Barr's University of Glasgow contract left him free to undertake whatever consultancy work he wished, he needed to keep the different domain of manufacturing clearly separated from his academic base.[86] In June 1895 he signed a lease for a 700 ft^2 (63 m^2) workshop in Byres Road, Glasgow, which was conveniently close and equidistant between the university and his home.[87] The move marked the start of a substantial increase in activity, but with a workforce of only six, including Harold Jackson and two boy workers, Barr & Stroud's Patents was still almost wholly dependent on out-sourcing for virtually all of its components and only equipped to do assembly work and some fine machining for experimental work like the Range and Order instruments.

At the same time, Barr began to put pressure on Armstrong's to produce some substantial business, demonstrating his growing awareness of naval and military affairs generally. At the end of April he warned Armstrong's that the move to Byres Road would cause short-term delays, but assured them that delivery times would subsequently improve. Five days later he suggested promoting the idea of rangefinders to shipping lines, and in the same letter asked if there was an opportunity to sell more to the Imperial Japanese Navy, whose Naval Attaché he had just met. In early May, he urged them to persuade the Chilean Navy to order rangefinders, and reminded them of their contractual obligation to 'influence prospective clients'.[88] In August, he badgered Armstrong's again, expressing 'disappointment' that no orders had yet come in. He reminded them that the Imperial Japanese Navy was ordering 'large quantities of new material' and hoped for 'an order for a considerable number' as a result, particularly as he had now provided Armstrong's with a rangefinder, free of charge, for demonstrations. Five weeks later he told his agent that he could not understand why foreign navies for whom they were building ships were not buying rangefinders, and suggested promoting them for land artillery as well. By November, Barr wanted pressure to be placed on the French and American governments, and then in the following January he pointed out that orders from Armstrong's were far less than the Admiralty's.[89] In fact, the armaments firm was not doing quite as badly as Barr implied, having sold fifteen rangefinders to seven different foreign powers since the agency was set up, but what he wanted was substantial orders, rather than the small trial purchases that were being made.

Armstrong's apparent lack of success with foreign clients resulted from circumstances that, ironically, the firm itself had largely created. The problem was not so much with the rangefinder, but with the question of what was to be done with it by its purchasers. There was still no consensus amongst tacticians to direct its use, not least because of Armstrong's success in promoting the 'quick-firing' (QF) gun which was one of their main ordnance specialities. The tactics of the QF gun prescribed large volumes of fire delivered rapidly at relatively short ranges, rather than deliberately aimed shots at greater distances. So long as Armstrong's were building warships whose main armament and selling point was the QF gun, the tactics of the weapon tended to diminish the usefulness of the rangefinder, which seemed more appropriate to the largest ships with the biggest and slowest-firing guns. Emphasizing this uncertain conceptualization of the rangefinder's use, the British Admiralty was as interested in the rangefinder's role for navigation and ship station-keeping as it was for gunnery control in the large warships where it had already been deployed.[90]

Despite Barr's frustrations, the business continued to grow. In 1896 the Admiralty ordered forty more FA2 rangefinders and foreign business added another fifteen to the total. The year saw another stage in the firm's enlargement, with seven extra staff taken on and, for the first time, some optical work being done in-house. It was indeed an important year for the firm. Barr made a lengthy trip to the US, ostensibly on university business but in reality largely as a research and marketing exercise for Barr & Stroud's Patents. This lasted from mid-April until late June and took in 'sixteen colleges and many engineering works' as well as opening up contacts with the US Army and Navy.[91] Apart from studying American engineering methods and business management practices, he gave quotations to the US Army's Chief of Ordnance and even got a US Navy order for a trial rangefinder. Barr came home convinced that, although there were sales opportunities in the US, if Barr & Stroud wanted to sell in quantity to the US government, then arrangements to manufacture there would be essential because of a prohibition on the purchase of war materiel abroad. He returned to a situation where, despite the growth, the business had developed some serious problems.

There were two particular difficulties retarding growth. First, the development of the electric Range and Order instruments which the Admiralty wanted to facilitate the rangefinder's use in gunnery control was bogged down, and second the problem of obtaining consistently high-grade optical work from the key supplier, Hilger & Co., was getting worse. Matters came to a head in January 1897, revealing serious tensions within Barr & Stroud's Patents. Shortly before then, Adam Hilger had raised with Barr the possibility of 'an amalgamation of some kind' that would benefit both firms.[92] That did not wholly appeal to Barr, who thought that nevertheless some kind of working agreement could be reached if Hilger moved part of his business to Glasgow and into vacant premises close to the Byres Road workshop. This, he told Stroud, would ease matters by avoiding the 'the great delays we now have in sending things back and forward and writing to and fro – just as we now have in writing about [Range and Order] recorders instead of talking the matter over with you on the spot'.

Stroud was still living and working 200 miles away in Leeds and Barr was frustrated by the difficulties this separation was causing. It was, he continued, 'a very serious matter' about the slow progress being made with the complex stepping motors and circuitry needed for the evolving control system.[93] It was 'quite impossible' to continue under current conditions, and 'the whole position requires to be well talked over and the course of the future mapped out'. Stroud should come to Glasgow without delay – 'Make some arrangement whereby you can come down' he concluded peremptorily. Three things drove Barr to lecture his partner so strongly and atypically. He saw the R&O system as crucial to the Royal Navy adopting the rangefinder on a large scale, and so opening an even larger foreign market. It was not a diversion from rangefinder manufacture, but an extension of it, so the delay in the process of invention was unacceptable. Jackson, upon whom Barr was increasingly relying, was 'quite down in the mouth' about the lack of progress and minded to leave; if he went, replacing him would be far from easy. And, for reasons that remain unclear, Stroud was loathe to visit Glasgow, a reluctance that meant every detail of design had to be sent by letter which resulted in misunderstandings and further delays because Jackson and Barr often found themselves dealing separately with him on aspects of the same problem. The tensions between the partners subsequently seem to have relaxed, although what remedies were taken remains unknown.[94]

The proposed association with Hilger did not go through, possibly because Stroud thought Hilger's standards would not automatically improve through moving to Glasgow, but more likely because Hilger's skilled workers were unwilling to go with him. Business also improved, although Barr continued to tell Armstrong's that they were not 'pushing' their clients sufficiently into buying them.[95] The workload increased enough to justify taking additional premises and the Admiralty ordered another fifty rangefinders before the end of 1897.

Most importantly, by June the following year, the R&O problem was finally solved and a viable system introduced which was offered to the Admiralty.[96]

Because the Admiralty had imposed no conditions of confidentiality on Barr & Stroud's development work, the partnership was able to offer the Range and Order system to foreign clients. In July 1898, before the Admiralty had made up its mind, the Imperial Japanese Navy took the initiative to install the equipment in every ship already fitted with Barr & Stroud rangefinders. This was the marketing breakthrough that Barr had been seeking. First, equipping a battleship with the installations, as well as rangefinders, more than doubled the value of business, adding approximately £800 to the £750 cost of the then typical outfit rangefinders.[97] Even more importantly, the system removed one of the main obstacles to persuading navies' gunnery specialists to adopt the rangefinder on a larger scale by providing an effective means to convey range readings around the ship, irrespective of weather or battle conditions.

Both foreign sales and the expectation of greater domestic business grew substantially during the second half of the 1890s. Between 1896 and the end of 1899, the Admiralty ordered 196 rangefinders and foreign navies 76, with sales receipts totalling £48,000 for the period.[98] Most importantly, the Royal Navy decided in 1898 to fit rangefinders on every capital ship in the war fleet, undoubtedly the best product endorsement that the growing business could have secured. In anticipation of the very large orders that would result from the Admiralty, and in response to the growing foreign interest, the firm looked for larger premises where the growing volume of trade could be better handled. In May 1899, Barr & Stroud moved into a factory building almost five times the size of the existing one, covering 3,360 ft^2 (302 m^2) in Ashton Lane, Glasgow, only 100 yds from the existing premises. The existing workforce was supplemented and new machinery was installed, thereby increasing the range of work that could be done and reducing the dependency on outside supplies of mechanical components, although the need to buy-in optical components was still not reduced. By late 1899, Barr & Stroud was running as the world's only 'naval rangefinder manufactory' with a workforce of about sixty, six of whom were university graduates working as a research and development team.

This first phase in the formation of the optical munitions industry was dominated by the armed forces' common recognition that developing military technologies had made the more extensive use of optical devices essential. It also saw the formation of what were to be greatly differing and persistent attitudes between the two services about the introduction of the principal innovation, the rangefinder. The Navy showed a greater willingness to engage with the possibilities it afforded and was much more prepared to commit itself to acquiring the new technology, even if its conceptions of how best to use it were still not fully formed. The emergence of the rangefinder and its manufacturing base were governed not just by the influence of advancing weapons technologies, but also by evolving notions of tactics and strategy that influenced the scale and variety of demand as well as the predicted use for optical devices in war. Where in the late 1880s there had been no identifiable optical munitions industry, by the end of the century there was a still small but already distinctive British manufacturing base for a device whose sole application was in warfare. During this formative phase, there was no experience of combat to act as an evaluator of the utility of the equipment being acquired, nor of the capacity of industry to deliver under pressure. The next part of the story examines the effect of warfare on the armed forces' attitudes to optical munitions, the business relationship between state and industry and the influence of developments in armaments technology on the interaction between manufacturers and the armed forces.

2 THE GROWTH IN IMPORTANCE FROM THE BOER WAR TO 1906

Before the start of the Second Boer War in 1899, the optical instruments employed by the British armed forces had yet to be used on active service in a major war. Although the British Army had been involved in fighting in colonial campaigns throughout the second half of the century, there had been no sustained experience of action to demonstrate the effectiveness of the equipment combined with the tactics into which they fitted or, equally importantly, the ability of the domestic industry to manufacture when faced with the increases in demand typified by large-scale and sustained warfare. By the end of 1906 the British Army had been able to digest how efficiently its optical munitions performed in the Boer War and the War Office had had the opportunity to consider how well its procurement processes had worked. The emergent optical munitions industry had also been given a taste of government contracting under the urgency and pressures of wartime conditions. The Royal Navy had seen no fighting at sea and thus had experienced none of the first-hand experiences of the Army, but it continued to be the industry's more important British service client, its demands increasingly driven by a combination of factors including evolving attitudes to gunnery, improvements in ordnance and the emergence of what amounted to entirely new weapon systems in the form of the submarine and the Dreadnought battleship. This chapter examines how British service attitudes to optical munitions evolved during this time and how they affected the growth of specialized instrument manufacture.

The War Office and Optical Munitions at War, 1899–1902

The Boer War was both the first conflict to see optical munitions employed to any significant extent and the first to trigger demands for the procurement of very large quantities of optical equipment at very short notice. This resulted not just from the scale and sustained nature of the fighting, but from the climatic and geographical conditions encountered in South Africa. In many ways, the topography provided ideal circumstances to demonstrate both the need for and the capabilities of military optics. Open terrain coupled with clear air provided almost constant unhindered visibility over distances that were far greater than troops had previously experienced or trained for. That combination maximized the potential for engagements at the long ranges made possible by the recently introduced types of small arms and artillery, whilst taxing the ability of troops to identify their targets quickly and engage them effectively.

The guns used in South Africa reflected recent progress in weapons technology and design. In particular, the new, more powerful nitroglycerine-based propellants which had displaced the less efficient black powder substantially extended the ranges of both small arms and artillery beyond those previously available.[1] The former, for instance, could now be used at ranges past 600 yds against individual targets and at more than twice that distance against large bodies of men. These capabilities found a perfect environment in South Africa, but required both clear target identification and precise aiming for fire to be delivered with accuracy and effect. Infantry and artillery weapons alike demanded an accurate knowledge of target distances once ranges extended past approximately 400 yds for the necessarily exact sight settings required. But even before sights could be set, it was necessary to locate and identify targets, a task made difficult in South Africa because they were often small numbers of men dressed in clothing which blended almost indistinguishably into the surrounding landscape. It became apparent once fighting started that the Army was less than well prepared either for identifying small and far distant targets or for measuring their range, and also that the War Office struggled to rectify those deficiencies.

For long-distance observation the Army relied on a variety of officially issued telescopes, all of which followed what had become a general pattern amongst their commercial manufacturers. Although, unlike small arms for instance, there was no standardized pattern in service, all the telescopes conformed to a basic contract specification which left the maker free to decide on the details of construction. Giving a magnification of twenty times, or '20x', the Army's telescopes were made up of a series of extensible concentric tubes which had a collapsed length of about 15 in. opening to an operating length of around 36 in. Weighing around 2.5 lbs, the telescope's tubes held a precisely worked optical system using specialized glasses in order to present the user with a sharp image. Such instruments were easily portable, reasonably robust, and powerful enough to identify man-sized objects in the landscape at distances of 2,500 yds. The great problem in South Africa was that there were simply not enough observation instruments for issue amongst either the infantry or the artillery and the Director of Army Contracts noted subsequently that the demand for more and better instruments began 'as soon as troops began to move over the country'.[2] The shortages continued and calls for better equipment were never properly met. A post-war survey of officers on the use of observation instruments during the war showed that although

only 9 per cent of respondents specifically criticized the quality of telescopes supplied, three times as many complained that the scale of issue was inadequate, even
by the end of the war.[3] The situation was even worse with binoculars. There, 13
per cent of replies reported dissatisfaction with one or more aspects of the issued
instruments, 22 per cent specifically preferred named commercial brands, and 40
per cent complained of insufficient numbers through the campaign. Shortages
mainly characterized the deployment of optical munitions during the Second
Boer War, followed by complaints of inadequate quality.

Similar comments seem not to have been formally sought on the performance
of the Mekometer which had begun to enter infantry service in 1891 and had
subsequently been adopted by the field artillery. That may have been because it
had enjoyed a singularly undistinguished war. Criticisms of its effectiveness had
already emerged when used on peacetime manoeuvres, and its eventual performance in action fell far short of expectations.[4] The topography and climate of the
veldt allowed Boer units regularly to open fire at long ranges, inflicting casualties
on British troops who were frequently exposed with little cover. Problems using
the Mekometer to find the range quickly for effective retaliatory fire soon became
apparent. The 1889 trials had been conducted in peacetime using clearly defined
targets under conditions which posed no threat to the operators' safety, but in
South Africa the enemy dressed to blend in with the background, making it hard
to identify a distinct rangefinding mark. Adding to that difficulty, Boer riflemen
were able to shoot at the Mekometer teams because they had to stand erect in the
open without benefit of cover or protection while taking readings.[5] Range readings produced by the instrument under these conditions were, understandably,
erratic and this undependability coupled with the ever present risk of high casu-
alty rates led to the instrument being little used in action.[6] However, even with its
use largely relegated to such mundane tasks as setting out lines of tents for camps,
there were still insufficient Mekometers to go round, and as with telescopes and
binoculars it too became a procurement problem for the War Office.

The War Office and its Dealing with Optical Contractors

The Second Boer War has been described as a notable 'failure in industrial mobilization' in the context of weapons and ammunition supply from both private
contractors and state factories.[7] Based on the contents of the War Office's own
post-war survey and analysis of the optical equipment in use, that understanding certainly applied to optical munitions, although there were no state-owned
manufactories. Adequate supplies of instruments were evidently lacking throughout the conflict. In the case of the interaction of the War Office and the British
optical industry, it has been suggested that the problems of procurement lay at
least partly in what has been alleged to be a symptom of a general 'malaise' in the

industry, reflecting a condition where, by the start of the war, 'the outlook for the optical industry seemed bleak'.[8] In this view, the war offered the opportunity for the optical manufacturers to reverse this trend by taking advantage not only of the resulting expansion of demand for familiar instruments such as telescopes and binoculars, but also new types such as the rangefinder. It may be tempting to see optical munitions as a basis on which to rejuvenate an industry which was allegedly on the verge of being moribund and already lagging greatly behind its expanding German counterpart,[9] but the evidence suggests a rather different situation.

The sudden surge in demand from the Army in South Africa certainly found the War Office with insufficient reserve stocks and highlighted the inadequate supply channels for many optical stores. As with artillery and ammunition, little thought had been given to ensure sufficient means rapidly to expand production in the event of war,[10] but with optical munitions there were other problems not found in increasing the output of field guns and cartridges. There was, quite simply, as yet virtually no infrastructure on which to build. Before 1899 the Army had employed only small numbers of observation and rangefinding instruments. Unlike rifles or field guns, telescopes were issued to regiments in small numbers – scarcely double figures. They had reached a design plateau by the 1870s and were still able to perform whatever tasks current tactics demanded before the British Army mobilized for action in South Africa, so that enforced obsolescence had not prompted any thoughts of large-scale replacement. Attrition rates had been low because, unless meeting with some accident, there was little scope for wear and deterioration. When additional ones were needed, they could be bought from any one of a number of makers, all of whom regularly made telescopes which conformed to the pattern required by the War Office. The binoculars, or 'field glasses' used by the Army were also little different from those made for the civilian market. They were simple instruments built on the 'galilean' pattern, using optical systems closely related to the telescopes and bodies of cast or moulded metals. Most of those being sold in Britain were actually imported from France where a substantial industry had come into being during the 1880s and 1890s, when interest in lightweight, easily carried instruments for casual observation of sporting events or nature had begun to grow.

But even if there were no regular relationships with makers of the kind that characterized the procurement of things like revolvers or explosives, it might well be asked why there should have been continuing shortages of relatively simple equipment which was needed in quantities which, compared to the Army's service rifle, were actually modest. There was never any question of issuing each soldier with a telescope or binocular, nor even to every non-commissioned officer, so why should delivery problems have persistently continued?

The answer was to be found in the War Office's procurement process and can be laid firmly at the door of the Director of Army Contracts whose shortcomings in handling the procurement of ammunition and weapons have been highlighted elsewhere.[11] However, in the case of optical munitions, the Army Contracts Directorate demonstrated a level of ineptness that seems to have exceeded the deficiencies it displayed in dealing with artillery and ammunition suppliers. There, it had been more used to maintaining harmonious relationships with its suppliers rather than pressuring them over deliveries which were in arrears, but with optical products it had no relationships on which to base its wartime practices. Instead of issuing contracts for the immediate delivery of established patterns of equipment which, for instance, happened with ammunition, the Contracts Directorate instead began issuing 'Requests for Tender' amongst the assortment of firms on its list of approved makers for optical or scientific instruments.

War Office procurement procedures were meant to ensure that the Army received products of suitable standards at prices which represented fair value to the Treasury. To be eligible to compete for War Office business, suppliers had first to qualify to be placed on a list of contractors who were deemed capable of manufacturing to whatever levels of quality were appropriate for the particular product. That qualification was obtained in the leisurely times of peace by supplying items for testing against the relevant criteria. Only after passing that test was a maker's name placed on the approved list for whatever category included that item. There were approved lists for every type of materiel which the Army used, including 'scientific instruments', a generic term which included everything from medical microscopes through to observation telescopes and all the other items of optical munitions. It was on the basis of that approved list that the Director of Contracts began to seek deliveries to meet the orders now coming in from South Africa.

There were two problems with the way the Contracts Directorate proceeded. Firstly, not having been in the regular practice of buying what was now needed and having no relationships to guide it, the department had no means of knowing which firms on its list had the requisite expertise or capacity to undertake the work expeditiously. In consequence it adopted the prophylactic policy of indiscriminately requesting firms to tender for what were seemingly haphazard assortments of equipment, irrespective of whether they had previously had any experience in producing those items.[12] Even as the war went on and the Contracts Directorate might have been expected to be learning from its experiences, that policy continued. As late as July 1901, a request to tender was sent to Barr & Stroud in Glasgow which amongst other things included the Mekometer and the Watkin Depression Rangefinder, neither of which the firm had ever made, when both were already being regularly manufactured by Thomas Cooke & Sons in York.[13] Also included were observation telescopes and telescopic sights for artillery, which the firm once again had never previously manufactured although a number of the larger London makers were already experienced in making them for the War Office.[14] Familiar with the products or not, Barr &

Stroud duly tendered; in August the War Office rejected its price for observation telescopes, despite the urgency of the need for them, but in September accepted bids for both the Mekometer and Depression Rangefinders. Quite why the latter instrument, intended for use only in coastal fortresses, was needed in the rush to provide equipment for the South African War remains unclear. Inconsistencies in the way orders were being placed is evidenced by a letter from the firm to the Director of Contracts soon afterwards saying that an order just offered for a hundred telescopic sights was 'not of sufficient magnitude' to be profitable at the Directorate's price. The following month Barr & Stroud wrote that it would not even tender for them in the smaller numbers then being requested. By February 1902, the company was clearly unhappy about doing business with the War Office, complaining that contracts were being issued for only parts of tenders, and objecting to the scaling-down or cancelling of orders already issued.[15]

Barr & Stroud's unwillingness to accept whatever terms the Contracts Directorate offered may have been the result of pique at the low value of War Office business that actually accrued to them during the period of the war; the total of orders resulting from the Boer War between 1901 and 1903 came to little more than £1,020 out of a turnover of almost £51,000.[16] But, nevertheless, the underlying theme of the correspondence was that Barr & Stroud was a business which existed solely for the manufacture of optical munitions and had both the capabilities and capacity to manufacture what the War Office needed. Those abilities were being under-used by the War Office partly because of the piecemeal way in which orders were allocated and partly because the Contracts Directorate had done little or nothing to discover what the firm was actually capable of doing.[17] If its dealings with Barr & Stroud were typical of its other dealings with the optical manufacturers, it difficult to avoid the conclusion that the Director of Contracts still seemed oblivious of how best to organize the distribution of orders even after three years' experience of attempting to meet the Army's wartime needs.

The actual total of optical munitions ordered by the War Office during the course of the war was relatively modest, certainly in view of the very large numbers of troops in South Africa. At its peak, some 300,000 men were in theatre there. To equip such a force, between April 1889 and the end of March 1902, just 978 artillery sighting telescopes were purchased, along with what were described as 3,842 'portable telescopes'.[18] A much larger number of binoculars was bought, totalling 10,255, most of which were the simple, low magnification galilean pattern.

The Annual Report of the Director of Contracts for 1901–2 noted that 'a considerable number' of the 5,810 binoculars bought that year were made 'on the continent',[19] and the next year that 'a considerable portion' of the 13,500 obtained was bought 'as usual, from the continent' at an average price of £1.25 each[20].

The Director of Contracts' summing up of his department's experiences after the end of the war noted that the demand for telescopes and binoculars was often 'as large as could be dealt with', something which he attributed partly to the result of past purchases having been 'small and to inferior patterns'. This was perhaps an instance of being economical with the truth in an attempt to put a gloss on the department's far from satisfactory performance in organizing its suppliers. Past purchases may indeed have been small because of the peacetime army's level of demand, but what had been bought earlier in the 1890s was essentially the same as was purchased during the war itself. The only optical product to appear in South Africa that was indisputably superior to any predecessor was the recently introduced prismatic binocular which provided both greater magnification and a wider field of view than the ubiquitous galilean pattern. The bulk of those purchased during the war had been bought in South Africa by the Army Ordnance Department and the rest by regimental commanding officers for issue directly in the theatre.[21] Reports from troops receiving these commercial types, which were mostly German made and on sale through local agents, showed how much better they were than the non-prismatic ones issued officially by the War Office. Only 10 per cent of those reporting on the latter were satisfied with them, whilst 95 per cent of those commenting on the prism patterns gave favourable reports. By 1902 these were already being made in Britain by at least two London companies but the Contracts Directorate made no purchases from them.

The benefit of War Office business to the optical manufacturers during the South African War was at best transitory. The immediate fillip to the trade given by increased military demand was not on a scale large enough to justify the creation of new businesses specifically for optical munitions, or even to set up special departments within existing companies. The increased demand created by war disappeared promptly once hostilities ceased and although the fighting had convincingly demonstrated the importance of optical devices in warfare it did not trigger any urgent subsequent requirements for entirely new patterns of optical munitions on a large scale. There was, however, an emerging willingness to look for improved equipment in the light of failings in some of the types used in the fighting, a willingness which offered a window of opportunity for optical firms to engage with munitions contracting.

The War Office and the Rangefinder Problem

The Mekometer had proved unsatisfactory in South Africa for both the artillery and the infantry units. Criticisms of it had been voiced as early as 1893 by no less a body than the army's School of Musketry,[22] and wartime experience had clearly shown the defects inherent in the two-observer type of instrument. Other alternatives had emerged in the 1890s, particularly those made by Barr & Stroud in Britain and by the German firm of Carl Zeiss at Jena.[23] Both of these were based on the self-contained, single-observer principle and because each firm offered them for commercial sale, officers serving in South Africa set about acquiring these instruments privately to replace the problematic Mekometer. Despite their high cost of around £50, officers who were unable to obtain them on loan from the makers were willing to purchase them outright. Their comments, published in the popular press as well as professional journals, invariably emphasized the superiority of these types over the Mekometer and other, older, service patterns which remained in service.[24]

Although both the artillery and the infantry now saw themselves badly in need of a reliable rangefinder, each branch of the service went its own way about finding a replacement. The field artillery continued for a time to be wedded to the idea of the long-base instrument, whilst the infantry began to reconsider what might best serve its needs.[25] In the artillery, the existing paradigm of the long-base two-observer rangefinders continued, despite the Mekometer's acknowledged deficiencies. Although having formed a favourable impression of the Barr & Stroud single-observer rangefinder in trials for fortress use in 1899,[26] the artillery not only ignored its potential as a mobile instrument but subsequently set out to 'reconsider the claims of the telemeter', a still earlier long-base device, simply because it was 'undoubtedly more accurate than the Mekometer',[27] which completely ignored the recent practical experience showing that the difficulties in operating a similar instrument outweighed any potential increase in accuracy.

The field artillery continued to deliberate about rangefinders in the three years after the Boer War, eventually deciding that the self-contained type offered overwhelming advantages. Even then, no decision on the large-scale adoption of the type was made until 1913. The infantry, although willing to start with the proverbial clean sheet of paper, eventually fared little better.

Unlike the gunners' leisurely process, the infantry's experience with the Mekometer was so bad that it set about finding a replacement even before the war was over. Less concerned with the last word in accuracy, the infantry turned more readily towards the demonstrated advantages of the single-observer rangefinder and in the autumn of 1902 began trials to select a new instrument.[28]

There were three principal contenders in the 1902 trials, two designs from outside the Army and one from a serving officer. One of the civilian designs was submitted by Professor George Forbes as a private venture, and the other by Barr & Stroud who, at the War Office's request, had prepared 'two specially constructed instruments' that were essentially smaller versions of the firm's now well-established naval models. The third was designed by Captain A. H. Marindin, an infantry officer who had been interested in rangefinders since 1895 and had produced his first working model in 1901, entirely at his own expense without any financial or technical assistance from the Army.[29] By 1902, Barr & Stroud's business with the Admiralty had provided them with rangefinder manufacturing expertise that was unmatched, not only in Britain but anywhere else in the world.

Having been specifically asked to submit for trials, the firm might be expected to have been confident about its chances of success, but its partners nevertheless had doubts about the outcome of dealings with the War Office. After the failure to get War Office orders in 1889, Barr had successfully turned the partnership's attention to naval rangefinders, writing off the War Office as a likely client largely because he and Stroud considered that any infantry model they made would be 'prohibitively expensive' compared to the Mekometer.[30] William Stroud particularly was convinced that the War Office was driven by price in considering optical devices, an impression that could hardly have been lessened by dealings with the Contracts Directorate in the Boer War. Early in the 1902 trials, Stroud was highly uncertain about the firm's chance of winning the competition

The reason for his disquiet was the rangefinder that Professor Forbes had submitted. It had come to Barr & Stroud's attention when an account of it was published in the journal *Nature* late in July 1901 and it seemed to Stroud to have features which made it a very strong competitor.[31] The device was quite unlike the Barr & Stroud rangefinder in using the principle of stereopsis (binocular or three-dimensional vision) to measure distances.[32] It was built around a prismatic binocular specially made for it by the German Zeiss company, to which was attached a folding lightweight accessory unit to provide the observer with images giving an enhanced degree of stereopsis. The binocular body's optical system provided what appeared to be a 'floating' ranging mark which the operator made to sit on the target by adjusting the instrument, so obtaining the distance reading. The Forbes instrument worried Stroud because its design offered light weight, potentially high levels of accuracy and, he believed, relatively low cost. Although Zeiss already made a patented rangefinder working on the same principle, Forbes's design did not clash with any of its patents and he had persuaded them to make the special prismatic binocular which formed the basis of his instrument. By late November, Stroud was so convinced of its advantages that he suggested taking up the idea because in his opinion Barr & Stroud could 'lick Forbes at his own game', seemingly because Forbes had been unable to patent those aspects of his invention which attached to the prism binocular. Furthermore, in Stroud's mind there was 'no justification' for making a short-base rangefinder of lesser accuracy.[33] His concerns, though, were dismissed immediately by Harold Jackson, now in substance if not in name the company's business manager, who reminded him that the firm had not been asked to design a rangefinder on a new principle, but to produce one on their established pattern: 'It is what the War Office has asked for', he wrote,[34] anxious to restrain Stroud from being diverted into potentially fruitless efforts that would do no more than inhibit the completion of the redesigned rangefinders for the competition.

Stroud's concerns would not have been lessened by the vigour and enthusiasm with which Forbes was promoting his device. Between November 1901 and the late summer of 1902 he presented papers on it and rangefinding in general to the Society of Arts, the Royal Society and the Royal United Service Institution. The final paper, delivered after his return, recounted his experiences and claimed wide endorsement by officers in the field, including the theatre commander, Lord Kitchener.[35] Forbes's promotional efforts were enhanced by his actually having spent time in South Africa, which further worried Stroud. He continued to badger Jackson, suggesting ways of beating Forbes and stressing the need to produce a rangefinder costing no more than the £25 expected to be Forbes's price, which was less than half that of the firm's own instrument, in order to 'win the day'.[36]

Although no one else at Barr & Stroud agreed with him and in the end the Forbes type was rejected, Stroud's concerns were by no means misplaced. But the real threat to his firm actually came from the third competitor in the trials, the Marindin rangefinder. That instrument was to deny the company the War Office's rangefinder business through a combination of factors that certainly included cost and probably embraced a high level of institutional bias.

Marindin's rangefinder worked on similar principles to Barr & Stroud's and he had approached the London firm of Adam Hilger & Co. Ltd in 1900 for help in producing a working model from his design.[37] Hilger & Co. were by then making the key optical components for Barr & Stroud's naval rangefinders, something which absorbed much of the firm's energies and absorbed most of the attention of its senior staff.[38] Hilger & Co.'s close involvement with Barr & Stroud meant the firm knew as much about rangefinders as anyone else in the country and was better placed to assist Marindin in developing his ideas than any other British company. Frank Twyman, then Hilger's manager and later its managing director, recognized the problems posed by the existing patents and translated Marindin's plans into a design which not only avoided the protected features of Barr & Stroud and Zeiss, but had sufficient novelty to patent in its own right.[39] Irrespective of Twyman's vital contribution, British Patent 16647/1901 was in Marindin's name alone, probably because Twyman was uncertain about Barr & Stroud's reaction to his involvement and unwilling to prejudice the somewhat volatile relationship between the two firms. Dealings between the two were not always harmonious, with Barr & Stroud regularly complaining about erratic quality and late deliveries, but Hilger's were by then tied into what had become a decidedly symbiotic, if somewhat asymmetrical, relationship. Hilger's relied on Barr & Stroud for a substantial part of its business and Twyman was unsure as to whether his Scottish clients would continue to use him as their preferred supplier if he tried to steer his company into rangefinder making on its own account.[40]

Whatever Twyman's concerns, by June 1901 Hilger & Co. was building a functioning rangefinder for Marindin.[41] Between then and June 1902, Hilger's submitted thirty-three invoices to him totalling £416, and by August 1902 had manufactured a number of prototypes, the first of which was sent to the Chief Inspector of Rangefinders as early as 20 December 1901.[42] Having experimented with them over the next seven months, and well before the start of the range-finder trials scheduled for the autumn, the Chief Inspector reported 'satisfactory results' to his superiors and on 26 August the War Office formally asked Marin-din 'to state on what terms he was prepared to offer his invention for the use of the Crown'.[43] This strongly suggests that the infantry officers taking part in the evaluation had already signalled their wish to adopt the Marindin as their stand-ard instrument and that the War Office was prepared to adopt it in preference to the other instruments scheduled for trial later in the year. None of this was known to either Forbes or Barr & Stroud, although the latter was certainly not over-optimistic as to the outcome of the trials.

In late September, George Forbes had sufficient confidence of success to propose to Archibald Barr that they should each concentrate on one of the Ser-vices.[44] Without revealing how he proposed to produce his design, he suggested that he should supply rangefinders to the Army, and that Barr & Stroud should withdraw from the competition and continue to supply the Admiralty. Barr's reply showed some caution as to the outcome of the tests. Having had dealings with committees at Woolwich before, he pointed out that there was no guar-antee that either of them would actually get any orders, and so any agreement would be premature. And he reminded Forbes that even if there were orders the financial benefits were uncertain as 'the War Office can claim the use of any patented invention with or without the consent of the inventor', implying, quite incorrectly, that the state had powers of sequestration without reward.[45] Under the circumstances, he saw no possibility of coming to any accommodation with Forbes and declined to go further.

Stroud's earlier pessimism was eventually justified, although Forbes was not the winner. In January 1903, the Rangefinder Committee reported its unanimous endorsement of the Marindin rangefinder's suitability, and its recommendation that 'at least 100 instruments should be provided for tests'.[46] No specific reasons for the choice were given. Barr & Stroud not unpredictably believed the choice was based on price, but Frank Twyman told the firm he thought it was because of its lighter weight.[47] Neither was apparently aware of the preference shown for it before the trials, and despite what he told Barr & Stroud, Twyman may actually have been surprised by the decision. He later noted that the 'government experts' on the Rangefinding Committee were convinced that the Marindin was not as well designed as any of its competitors, an opinion with which he privately agreed, despite his role as its manufacturer.[48] Its imperfections may have resulted in the lack of any substantial order in 1903, despite the Committee's recom-mendation. The Army asked Marindin for further trial models in 1903, and then carried out more rangefinder testing in October 1904, February 1905 and July

1906, each time asking for modifications to his design.[49] None were ordered for troop trials, and a final decision to take it into general service was not made until early 1907, an account of which will be given in the succeeding chapter.

The Admiralty and the Optical Munitions Makers

Although the War Office did little to stimulate the growth of the optical munitions industry in this period, the Admiralty was both more forward-looking and generally more technologically minded. In 1899 the Royal Navy was already a customer for complex optical munitions, even though it still employed them on a relatively small scale. Up to then it had bought 123 Barr & Stroud rangefinders out of the 189 the firm had sold,[50] but in the next seven years its purchases of rangefinders and other sophisticated optical instruments grew substantially as more attention was paid to gunnery at increasingly long ranges and with the arrival of the submarine as a potential weapon of war. By the time that the lessons of naval warfare in the Russo-Japanese war of 1904–5 were being digested the Admiralty had become a substantial buyer of complex optical devices.

The Aiming Problem and its Encouragement of the Industry

Even without recent experience of war at sea, the Royal Navy was aware that aiming its guns properly could be problematic. There were even difficulties hitting stationary targets at distances that were far less than the ballistic capabilities of the latest ordnance. In 1900, firing tests against a battleship moored at 1,700 yds, or approximately 1,500 m, showed that more than 60 per cent of the shots missed.[51] Two reasons could account for this poor showing. Setting the range wrongly on the sights and/or not aiming the guns properly at the target could cause projectiles to miss their targets. Setting the range correctly was vitally important. Admiralty ballistic tables showed that at 1,700 yds, in order to hit a 20-ft-high target representing a ship, the permissible range setting error was 142 yds. At 2,000 yds it fell to only a quarter of that, and at 3,000 yds was only 24 yds. Having an accurate range was not the only essential element. Aiming correctly was equally important but the 'open sights' in general use required the gun-layer to line up two widely separated points on the sighting mechanism with the target itself, giving considerable scope for human error.[52] Unless all three were correctly aligned, the projectile was unlikely to find its target. An aiming error coupled with a range-setting error was a certain recipe for failure and could cause a target the size of a battleship to be missed even at what were, in the context of the latest naval ordnance, very short ranges.

By 1900, although the question of measuring ranges was being addressed by the progressive introduction of the Barr & Stroud rangefinder, the problem of improving gun aiming was still to be settled. Optical sights offering the twin advantages of a magnified image and a single aiming mark in the same optical plane as the target had actually been in service since 1887, but their practicality had never been thoroughly tried and their employment had even been discouraged by an Admiralty Order in March 1896.[53] Aiming problems were tackled robustly after 1898 by Captain Percy Scott, who became an eloquent and aggressive advocate for the use of telescopic sights.[54] Scott not only revived their use but also became a ruthless critic of the quality of the existing types in service, lobbying for more powerful types with finer aiming marks.[55] His appointment to command the gunnery school, HMS *Excellent*, in April 1903, gave him the opportunity for still more effective advocacy of improved types and led directly to the Admiralty's decision in 1905 to carry out a wholesale revision of gun sighting and gunnery control arrangements of all fighting ships in the fleet.

In 1904, HMS *Excellent* had prepared a report on the Navy's sighting equipment which recommended the general introduction of new improved telescopic sights.[56] On 11 May 1905, the First Sea Lord approved a programme to accomplish the 're-sighting' of the entire fleet, a large and costly project to be funded out of both the current and following years' Naval Estimates. A circular in June showed the extent of the proposals, detailing which ships were to receive what telescopes, and pointed out that the changes would be phased in gradually in view of the quantity of sights required.[57] The 1905–7 programme represented, numerically, the largest order for optical instruments that the Admiralty had ever placed, amounting to approximately 4,000 assorted telescopic sights of new design.[58] The Director of Naval Ordnance hoped that existing telescopes could be used in the programme if found suitable, but in the event both of the patterns of sighting telescopes then in service were discarded.[59]

Five new types were to be ordered for distribution amongst different calibres of guns and their mountings. Two were of fixed magnification at 3x and 6x, and three of variable magnifications at 3x to 9x, 5x to 15x, and 7x to 21x. These were undeniably specialized optical munitions items rather than adaptations of civil instruments. Their optical systems incorporated specially arranged aiming marks known as reticles, and were housed in robust non-ferrous bodies intended to withstand the rigours of sea service and the shock sustained when guns were fired. The general specification governing magnification, angle of view and the necessary connections for attaching them to the gun mountings were strictly defined, but the contractors were left free to decide for themselves how to design the optical systems.[60]

The orders were to be divided between two established London makers, Ottway & Co. Ltd and the Ross Optical Co. Ltd, both of whom had previous experience in supplying various types of telescope to the Admiralty. Ottway received orders for all five patterns, Ross for only two. Using the values given in the Admiralty's *Rate Book for Naval Stores*,[61] it is possible to assess the total contract value as about £50,000, spread over the financial years 1905–6 and 1906–7. This was indeed a significant order – approximately £5.2 million at 2011 values.[62] Some measure of its size and importance to the optical munitions industry can be gained from comparison with the approximately £22,000 of rangefinder orders Barr & Stroud received from the Admiralty in the same years.[63] In the absence of company records for either Ottway or Ross, the effects this business had on them must be conjectural, but Ross's advertisement in the 1907 edition of *Jane's Fighting Ships* announced that they had made extensive additions to their works in consequence of what they demurely described as 'increased demand' for their telescopes, and that their production and prices would benefit as a result. For one firm at least, optical munitions work and the patronage of the Admiralty had a recognizable commercial spin-off.

Important as this business was, it was a one-off transaction that was unlikely to be repeated in the foreseeable future. The Royal Navy's very size, and, it may be argued, its earlier backwardness in failing to keep up with the growing potential of naval gunnery, provided a unique business opportunity for the firms who won sighting telescope contracts in 1905. Once the re-equipment was complete, demand for sighting telescopes would revert to being geared to new shipbuilding and replacement of losses through attrition. The telescope as an instrument offered little possibility of radical improvement in design or performance, so early obsolescence was unlikely, and although the 're-sighting of the fleet' was a costly operation, the telescopes themselves were not individually of particularly high value. The rangefinder, though, was a much more complex and expensive device that was still evolving, and new, improved designs had the potential to render obsolete earlier versions, creating a self-sustaining demand. Welcome as the sighting telescope orders undoubtedly were, they did not presage the development of a sustainable new branch of optical munitions manufacture. Their significance was that, apart from boosting income in the short-term, they firmly established Ottway and Ross as the Navy's telescopic sight makers, a status which was sustained by the shipbuilding programmes that continued until the First World War.

At the same time that new sighting apparatus was being considered, the construction of the novel battleship *Dreadnought* emphasized the pressing need for a rangefinder of greater accuracy over longer ranges. Unlike earlier capital ships, *Dreadnought* had a main armament of uniform calibre where five turrets, each mounting two 12-in. guns, replaced a mixture of turrets with guns of different calibres. The guns themselves were little different from those of immediately preceding battleships, but the important difference was the potential improvement in the damage that the new ship's heavy armament could inflict at longer ranges. To hit at those increased distances demanded greater precision in the aiming process. Errors in aiming and distance setting had to be eliminated before satisfactory shooting could be expected. The whole question of 'fire control', the integration of all the problems involved in long-range artillery fire at sea, began to be studied seriously in late 1903 and 1904, even before the design of *Dreadnought* had been finalized. Once again, as in the previous decade, technological advances were so strong that they not only challenged established norms but demanded investigation of the way to further efficiencies in gunnery. The Royal Navy was unavoidably faced with the need to provide a targeting system that would enable an unguided projectile fired from one moving ship to hit another moving vessel whose course between the projectile's despatch and arrival was unpredictable. The start of the process had to be the knowledge of the range of the target vessel, and as any error in range would disrupt the possibility of accuracy, the performance of the rangefinder was of paramount importance. Even if the rangefinder could not guarantee a solution to gunnery direction, without it control would be inadequate and the performance of the entire weapon system that was the battleship would be devalued.

Barr & Stroud: Emergence as the Principal Optical Munitions Maker

The increasing commitment to long-range gunnery, and the concomitant necessity for fire control, meant that Barr & Stroud was virtually guaranteed a monopoly of Admiralty rangefinder business by 1905. No other British optical maker had that firm's accrued expertise in the optics or mechanics of rangefinder construction, nor such an established working relationship with the Royal Navy. That relationship was, and continued to be, governed as much by the interaction of Barr & Stroud's senior personnel with the Royal Navy's gunnery branch and the procurement policies laid down within the Admiralty as it was with technical change. It was the Royal Navy's willingness to accept Barr & Stroud as its monopoly supplier as much as the firm's command of technology that allowed the firm to build up not just its domestic business but an even more successful export trade up to the close of 1906.

The Admiralty had indicated during 1898 that it wanted to acquire a very substantial number of rangefinders and because of concerns that prices were excessively high through Barr & Stroud's monopoly, had raised the question of acquiring the rights to produce them, either manufacturing the rangefinders itself or using other contractors.[64] The firm's response indicated how far it had shifted from the founding partners' original intention to derive an income from licensing their patents to others. Having had the potential monetary value of the rangefinder indicated by the growing interest of the Admiralty and by foreign enquiries, Barr & Stroud was now much more interested in supplying than in licensing, and had no inclination willingly to relinquish its monopoly. At the end of May 1898, the firm reiterated its insistence on a royalty of £100 per instrument made by anyone else, and insisted that the selling price of £250 for each rangefinder was absolutely the lowest possible. The firm refuted robustly the Admiralty's allegation that the bulk of the selling price represented 'royalty and commission', rather than a more usual mark-up on manufacturing costs.[65] The company's riposte was that, besides materials costs, the final price actually reflected the expenses of setting up, research and development, and a return on the accrual of expertise. To this aggregation they added a percentage to cover labour costs, operating overheads and then a final margin for profit.

Faced with what may be interpreted as either a reasonable commercial assessment of their products' value, or as downright obduracy by the company, the Admiralty abandoned the idea of acquiring the manufacturing rights and eventually issued a contract for 100 rangefinders at the end of June 1899, at the price demanded by Barr & Stroud. The Admiralty found itself in an unusual position with Barr & Stroud. Despite being the sole domestic customer, the Admiralty was never able to exert its 'enormous market powers' over Barr & Stroud.[66] Even though the Patents Act gave the Crown the right to 'use [a patented] invention for the service of the Crown' without the prior arrangement of terms or conditions,[67] no matter how much the Admiralty may have objected to the firm's prices or the proposed licensing fee, it was hardly in a position to take advantage of that right. The Act did not give the Crown any power to compel an inventor to manufacture for it, nor was there any other maker of naval rangefinders to whom the Admiralty could turn in the hope of obtaining a better deal. The scale of its demand had been insufficient to support more than one maker, even if any other firm had actually been able to produce competitive instruments. Barr & Stroud's various patents made the task very difficult, and the key one covering the prism and range-scale arrangements to measure and display the range remained in force until 1903.[68] The only potential rival for Barr & Stroud in 1898 was the German Zeiss rangefinder, which was built on a fundamentally different principle and itself well protected by patents. The Zeiss instrument was both foreign, which made it less than desirable to the Admiralty, and exceedingly demanding in manufacture, which would probably have made it even more expensive than the Barr & Stroud instrument.

The Admiralty's options were limited either to agreeing to the company's royalty demands or to paying the price demanded for complete instruments. The decision to continue buying from Barr & Stroud rather than seeking another maker was doubtless influenced by the company's emphasis in its riposte on setting-up costs and accrued expertise. If the Admiralty wanted another, and presumably cheaper, source for its rangefinders, it would need a contractor both willing and able to manufacture at a lower price, and would also have to allow for delays while such a firm became proficient in making a specialized instrument which was quite unlike anything else being made by the British optical or scientific instrument industry. No doubt these difficulties persuaded the Admiralty to maintain the status quo, a circumstance that repeated itself some fifteen years later and will be described in due course. However, the continued dependence of Barr & Stroud on outside suppliers for many of its components was a potential weakness for the firm that might have provided a means by which another source of supply could have been established had the Admiralty been so minded. So long as the patents remained in force, Barr & Stroud could dictate royalty terms, but once these expired in 1903, then the firm might be vulnerable to competition, particularly if it was still dependent on external contractors. The Admiralty's principal need in 1899 was the immediate acquisition of rangefinders, and short-term priorities overcame any question of future alternative sources of supply. The company was therefore safely able to anticipate the Admiralty's new business and proceed with plans to expand both the scale and scope of their operations.

By the time the contract was signed in mid-1899, Barr & Stroud had completed its expansion into a purpose-built engineering workshop equipped with a range of machine tools to allow the manufacture of at least some of the components used in the rangefinder, its mounting and the Range and Order instruments which formed part of the rangefinder's shipboard operating system.[69] This investment was in a year when turnover declined from over £14,000 to £8,500, and it represented the anticipation of Admiralty business rather than a response to orders already received, showing the measure of confidence which the firm was starting to show in itself. The move marked the beginning of a period of sustained expansion which was to lead the firm into a second move only three years later. This short interval saw the company's business expand and diversify, not just in the products being made but in the clients to whom they were supplied. This growth also directed Barr & Stroud towards an increasing level of autarky which was achieved only with some difficulty.

The period between 1899 and 1903 was not without problems for the firm. Turnover shrank from £27,731 in 1900 (largely composed of receipts from the Admiralty's large order of June 1899) to £15,070 in 1901 and to £14,522 in 1902.[70] Even with the Admiralty's decision to order rangefinders in quantity in 1899, William Stroud began to doubt the long-term viability of munitions work. In that year he wrote a series of letters from Leeds to Archibald Barr revealing an assessment of the firm's prospects which was much less optimistic than his partner's views.[71] He was still living in Yorkshire and working as Cavendish Professor of Physics at what had then become the University of Leeds, very rarely visiting Glasgow. In March, when in poor health, he wrote a letter revealing his thinking. Referring to the drafting of a new co-partnership agreement to avoid the possibility of bankruptcy following the death of one partner, he wrote about the need for diversification. He had already touched on this and the firm had designed a vacuum pump which had yet to achieve any useful sales. The continued lack of any alternative to rangefinders caused him to write that 'I regard the business as a very precarious one. If we had pumps really selling, and recorders [Range and Order instruments] &c &c I should believe in the stability of B&S much more'.[72]

Between April and July, still in poor health, he wrote a series of generally pessimistic letters on the financial problems that might result from the death of a partner but then, probably recovering from his illness, he became more positive. He specifically urged on Barr the need to reduce manufacturing costs so that, if necessary, selling prices could be reduced when the original patents expired in 1903 and other competitors might appear.[73] The point about the likely problems when the original patents expired was particularly telling, and ties in with the decisions made soon afterwards to expand and assume greater control of components manufacture. But despite Stroud's concerns about the future prospects for military and naval orders, the business was actually starting to grow substantially. The subsequent decision in 1902 to move again, this time to a much larger site, was taken because of the developments in naval business, and not through any programme of diversification into non-military products. The revenue from the firm's own vacuum pumps and an electric clock made under licence was very small indeed.[74] The order records after December 1900 show little demand for them, their individual selling prices were less than a tenth of a rangefinder and their contribution to the business could at best have been only marginal compared to that from naval orders.

The value of orders received for rangefinders and associated items grew steadily in value after 1901. This came partly from the Admiralty's increasing investment in fire control instruments. The original specification for the rangefinder in 1891 had called for the provision to relay ranges from the instrument to the guns, and Barr & Stroud had begun work on such apparatus at the same time as the rangefinder. In 1892 the Admiralty had decided that there was no immediate need for this transmission device, but the firm carried on and by 1893 they had developed a basic design which was submitted for trials in 1894.[75] Having announced that the system would be installed, the Admiralty continued testing until 1901, with the design of the Range and Order instruments evolving steadily. In that year, despite Barr's irritation at the time taken, the first of a series of substantial orders was placed.[76] Between 1901 and the end of 1906, British contracts for fire control instruments totalled £33,522 out of orders totalling £93,563.[77]

Important though it was, the Admiralty business was overshadowed by the growth in overseas orders in the same period. In every year from 1901 to 1906 foreign orders were greater than domestic ones, adding up to £149,569. The customer records show that these orders were almost entirely for rangefinders, the reverse of the pattern of Admiralty ones, implying either that foreign navies had failed to appreciate the need to integrate rangefinding into a gunnery control system, or that the extensive and expensive shipboard modifications needed to accommodate the electrical circuitry were unacceptable. Only the Imperial Japanese Navy was a regular purchaser of control instrumentation, but by no means on the same scale as the Royal Navy. In 1904, when Russian and Japanese purchasing was its greatest, less than 5 per cent of the spending was on fire control apparatus.[78]

The largest growth in rangefinder business began after the introduction of an improved model, the FA3.[79] There certainly was an increase in Admiralty orders for rangefinders after 1903, but whether the FA3 itself was responsible for them is not certain. By 1904, the Royal Navy was increasingly accepting that gunnery improvements were possible through the better methods emphatically prescribed by Captain Percy Scott when he took charge of the Navy's gunnery school, HMS *Excellent*, in 1903,[80] and it is as likely that the resulting willingness of the Royal Navy to entertain a revision in its approach to gunnery caused larger purchases of rangefinders, rather than simply the availability of a better instrument. As for foreign business, the escalating tension between Russia and Japan would have generated the same orders, irrespective of recent technical advances. The Russian Navy was severely disadvantaged in its gunnery methods compared to the Japanese and its purchases of the older and less expensive FA models suggested that it was principally concerned with quantity rather than the latest improvements.[81]

Even before this surge of business, the actual and expected growth in orders for an increasing range of naval gunnery instruments led in 1902 to the decision to build a much larger factory.[82] The existing site in the crowded West End of Glasgow was unsuitable for expansion because of the proximity of surrounding buildings, and the workshops themselves, spread over three levels, were increasingly inconvenient and not big enough to handle the larger instruments now being considered for development. The new site at Anniesland was some two miles from the existing works, in largely open country, adjacent to a railway station and at the end of a tram route which conveniently served the areas where most of the existing workforce lived and from which extra workers might be drawn. The firm hoped the clear air of the more rural setting, free from the effects of Glasgow's atmospheric pollution, would allow the final visual checking of the rangefinders to be done more efficiently and without interruption. A further benefit, from Barr's point of view, was that the new factory would be 'a place where workers could earn a decent living under clean, healthy and happy conditions'.[83] The land for the new works was purchased in 1902, and building began in the autumn of 1903. Additions to building plans were made in 1904 and 1906, largely in the expectation of more Admiralty business. The story of Barr & Stroud between 1899 and 1906 is largely one of expansion and profitability, based partly on the firm's own abilities and partly on a fortuitous combination of circumstances which saw a steady increase in concern with gunnery in the Royal Navy, bolstered by the profits generated from supplying both protagonists in the Russo-Japanese war.

Some organizational weaknesses became apparent in the development of Barr & Stroud after 1898. When the firm moved into the Ashton Lane factory, its chief abilities were in mechanical and electrical engineering, rather than in optics. Archibald Barr's own abilities lay firmly in those fields, and the staff he had recruited in Glasgow added to this strength. William Stroud, who was still living in Leeds, was the only person in the firm able to design optical components, and he also played a very significant role in the design of the Range and Order instruments which were to assume an increasingly important part of the business after 1901. Stroud's location, over 200 miles away, was to cause difficulties in the process of product development.

Stroud's letters to Glasgow show the volume and detail of correspondence between him and Archibald Barr when new designs were in progress. Daily letters suggested ways to overcome difficulties, and arguments over the best ways to proceed were conducted on paper. At times, Barr's frustrations were evident. In 1904, when the War Office had asked for a design for a new type of artillery sight, Stroud had dismissed the instrument as being of no value. Barr wrote to him bluntly, saying 'I do not agree with you, but we need not discuss that; they are wanted and are to be introduced into the service'[84] In December 1904, Barr wrote 'I do not agree that you can do the best for B&S by staying at home [in Leeds]'.[85] Stroud obdurately refused even to visit Glasgow and his continual, and seemingly determined, absence could hardly have helped the process of optical design development.[86]

In 1899, Barr & Stroud relied on one principal supplier of mechanical components and two suppliers of optical parts. James White, the Glasgow firm with which Lord Kelvin (Barr's earlier mentor at the University of Glasgow) was closely associated,[87] supplied castings and fabricated parts for both the rangefinder and its mounting. Chadburn Brothers of Sheffield, Yorkshire, supplied some spherical lenses and plane glass panels, and Adam Hilger of Camden, London, made the quality-critical spherical lenses for the main optical part of the rangefinder as well as all the prisms used in it. Complaints about the mechanical components seem to have been infrequent, but there were constant difficulties with the quality of the optical contractors' products, as well as their delivery times.

The problems with optical components were frequent and sometimes serious. Chadburn Brothers had been suppliers of the simpler optical parts since 1889, when they made them for the very first instrument.[88] Barr & Stroud used them for the less critical components in the rangefinder's optical system, such as the optics for the aiming viewfinder and the protective glass covers for the objectives. Even with these relatively simple items Barr & Stroud frequently returned parts to Chadburn's with complaints about inadequate quality and errors in execution,[89] but rather than looking for a replacement supplier, the company seemed content to instruct and educate, presumably because there was no other closer or more convenient source.[90]

The relationship with Adam Hilger & Co. was particularly important because for much of this period it was practically the only company in Britain able to provide the most important optical parts of the rangefinder. The Barr & Stroud instrument used two telescopes whose objective lenses were widely separated and provided the operator with an image of his target which was divided horizontally into two segments: these images were displaced relative to each other, but could be brought into alignment through a system of prisms to provide a direct reading of the target's range. The telescopes were not particularly complex in design, but it was important that they provided images of identical magnification in order to let the operator align them precisely and so obtain an accurate result. They presented a manufacturing problem rather than a design difficulty. Stroud specified their necessary magnifying power and angle of view required through the telescope system and left Hilger & Co. to select appropriate optical glasses and to compute the lens curves necessary to provide what was needed. Stroud's particular skill lay in the design of the complex prism systems which provided for the superimposition of the telescopes' images, but as with the telescopes he was entirely dependent on Hilger & Co. to produce them accurately; not achieving precisely the specified angles would cause a prism to fail in its purpose.[91]

Problems with both quality and delivery times from Hilger's had been evident almost from the start of the companies' relationship. Sometimes its work was praised, but often it fell below Barr & Stroud's requirements. As early as February 1893, Barr & Stroud were returning prisms as unsatisfactory and difficulties continued regularly.[92] In 1897 Adam Hilger, presumably as a result of the increasing volume of orders, mooted the idea of moving the business to Glasgow, and Barr reported to William Stroud that 'Hilger appears to favour an amalgamation of some kind'.[93] However, despite the delays caused by sending things back and forward, neither partner was enthusiastic about it and nothing came of the idea. The following year, however, Barr & Stroud complained that the defects in Hilger's prisms were causing them 'endless worry and expense'.[94] Between October and December of 1900 a series of letters to Hilger written by Harold Jackson, Barr & Stroud's general manager, showed how bad matters had become between the two firms.[95] The correspondence also highlights the problems of producing optical components to very fine tolerances.

Having had yet more problems with prisms, Jackson warned Hilger & Co. in early November 1900 that Barr & Stroud now had the means to check precisely the standards of optical work delivered. Towards the end of the month, Jackson was saying that Hilger & Co.'s proposed mutually acceptable standards were 'ridiculous' and on 30 November he threatened to go elsewhere for prism work. The threat, which was repeated only four days later, was really a hollow one, because Barr & Stroud had no other source to turn to. Its real purpose was doubtless to encourage Hilger's to improve its quality, but Jackson was overzealous and succeeded only in pushing the London firm to a point where it wanted to cease doing optical work for Barr & Stroud. A complete breakdown in relations between the two firms during December 1900 was averted only by Barr's personal intervention in a letter of 13 December, apologizing for the 'hurt' which earlier correspondence had caused.[96] But even this conciliatory letter reiterated (if less harshly) the possibility of taking orders elsewhere, and difficulties between the firms over quality control continued to surface periodically, although business between them continued without interruption.

The dependence on Hilger clearly concerned Barr & Stroud, and they periodically investigated obtaining optical components not just from other domestic sources, but from the German optical industry as well.[97] In 1897 they had made C. P. Goerz of Berlin their German agent, and afterwards periodically bought lens samples from this company, as well as asking for quotations for the manufacture of prisms.[98] In 1899, they attempted to buy objective lenses from Steinheil of Munich, but encountered problems similar to those they had already had with Hilger: either the very specific instructions given were not adhered to or the quality was inadequate, and even when satisfactory, the price was frequently considered excessive.[99] The same year, Stroud suggested asking Carl Zeiss of Jena to quote for optical components, doubtless because Zeiss had achieved the reputation of producing the best possible quality in optics. Sound as his suggestion might have been, there is no record that his idea was followed up.[100]

Barr & Stroud's continued dependence on Adam Hilger through to the end of 1906 was principally the result of an established relationship that, despite frequent problems, worked well enough to let them produce satisfactory rangefinders. It was also because they still lacked the expertise to do the work themselves, although after 1904 they began to organize the means to do this, partly from the desire to control costs but also to gain greater control of the quality and speed of delivery of components for experimental work.[101] Attempts to obtain more satisfactory quality from German suppliers had failed, suggesting that contemporary perceptions of the superiority of the German optical industry were by no means justified, and even if Barr & Stroud had been inclined to work their way through the entire catalogue of German makers, the growing inclination of the Admiralty to be independent of foreign suppliers even for materials in British-made products would have been a strong deterrent. Given the lack of any other British firm who could be relied on to perform better than Hilger's, Barr & Stroud had little alternative to becoming optical workers themselves. This process began in 1904, but only developed significantly after 1907.

The Submarine Periscope and Sir Howard Grubb & Co. Ltd

If optical munitions enabled capital ships to become more efficient fighting machines, they had an even more fundamental influence on another naval vessel. Without the periscope, the submarine would never have become a viable type of warship.

In 1901, the Royal Navy acquired its first submarines to evaluate the menace posed by the underwater vessel armed with torpedoes, and to determine the best ways to counter it.[102] The threat of the underwater vessel lay principally in its invisibility when submerged, but to exploit its potential the vessel's crew needed to be able to see what was happening above the water in order to navigate and position the boat for an attack. Neither the idea of the submarine nor a device to see from it was new; experimental vessels had been built by several navies in the late nineteenth century, and all used some kind of primitive device to permit observation when under water.[103] The effectiveness of these early methods was far from satisfactory, many being little more than glazed panels in an extension of the boat's hull that projected above the water when the vessel was below the surface. In other cases, combinations of simple lenses and mirrors were employed in a tube passing from the crew space through the hull to reach above the water. These 'periscopes' were more useful, but by no means widely adopted in the early submarines partly because of their optical limitations and partly because of the mechanical problems of making them watertight and durable. The early development of the submarine as a weapon was inhibited as much by the lack of the means to see as by any other engineering difficulties.

All the first British submarines had periscopes that were, by contemporary standards, effective enough to allow the boat to be used as a weapon. The earliest ones were made up to the specification of Captain Reginald Bacon in 1901 or 1902.[104] Bacon, a leading proponent of the military utility of the submarine, was subsequently introduced by the boats' builders, Vickers, to Sir Howard Grubb, the owner of the Dublin astronomical telescope making firm, who, according to Bacon, subsequently produced an improved version of his original design. [105] Bacon's claim may have been mistaken as Grubb's first periscope patent was granted in 1901.[106] The patent specification shows this to have been a sophisticated prismatic design, providing an erect, normal image, unlike earlier devices which either reversed or inverted what the observer saw. Grubb's good relations with Vickers gave him a monopoly of supply for all the Vickers' submarines built in the next five years, and he would have the vast majority of periscope business from the Royal Navy until 1914. Grubb may even have supplied periscopes for the US submarines built by the Electric Boat Co. as early as 1902, and was certainly one of the major manufacturers in the early years of the submarine adoption.[107]

The subsequent fostering of the periscope industry was different to range-finders and, so far as can now be seen, the Admiralty did not originally lay down any particular optical characteristics or qualities, nor did it ever liaise closely or even directly with Grubb. The numbers needed were small, only one for the earliest boats and two each in those built after 1905 and, unlike rangefinders, the shipbuilders delivered the submarines fitted with periscopes ordered by them and not by the Admiralty. Even if the monetary value of orders was still small, the importance of the periscope in the development of both the submarine and the optical munitions sector cannot be underestimated. As the rangefinder made effective gunnery at long ranges possible, so the periscope permitted the submarine to become a viable fighting ship. It became not just a navigational tool, its original purpose, but also the sighting device to permit the submarine's offensive weapon, the torpedo, to be aimed with precision. As the submarine evolved into a viable weapons system the presence of a domestic manufacturer became increasingly important to the Admiralty.

The period from 1899 to 1906 saw an increasing diversity in the industry and a growing, if not always satisfactory, relationship with the War Office and the Admiralty. For the War Office the experience of war highlighted weaknesses in both its equipment and its methods of acquiring it, although the lessons derived were by no means quickly taken to heart. As the Royal Navy had begun to display an increasing realization of the potential effectiveness of long-range gunnery and new types of ship, the Admiralty recognized that optical equipment would increasingly be vital for the effective deployment of sea power and began to lean more heavily on its acquisition. At the close of 1906, the optical munitions industry in Britain was both larger and more important than at the start of the Boer War and its dominant member was establishing an export business substantially larger than the market provided by the British state. The next phase in the industry's progression would be a massive uptake in demand both at home and abroad that was fuelled by what amounted to the general rearmament in Europe and beyond.

3 EXPANSION AND CONSOLIDATION, 1907–14

By 1914 armies and navies had become dependent on optical devices for much, and sometimes all, of their ability to use their weaponry effectively. That was particularly true for the major naval powers whose battleships and submarines were practically impotent without their rangefinders, telescopic gun sights and periscopes. Land forces were not so totally reliant but, even so, all deployed optics on an increasing scale and would have been hard pressed to counter an enemy in their absence. The seven years running up to the start of the First World War saw optical munitions production grow at an increasing rate and by 1914 a clearly identifiable sector of industry was engaged permanently in the production of such instruments which, with few exceptions, had no civil applications.

Only a small part of the optical instruments trade was engaged in this work, reflecting not just the specialized nature of what was being made but also the contemporary scale of demand for military and naval optics. That demand grew after 1907 partly because advances in optical technology permitted the creation of new instruments but even more because developments in weapons technologies and increasing political instability created a climate that encouraged European states in particular to increase their expenditure on armaments and take up equipment which increasingly depended on optical instrumentation for its effectiveness. For the first time, the British War Office became a systematic buyer of optical munitions, greatly increasing its spending in the last two years of peace. Even though its budgets for such equipment were far less than the Admiralty's, it began to bring firms routinely into munitions work and established them as regular contractors. Although the Royal Navy's demands had increased at a faster rate than the Army's, it employed a smaller number of firms to produce its requirements and established closer working relations with them than the War Office did with its contractors. This chapter examines the extent to which the optical munitions makers benefited from government business, assesses conceptions of the industry at this time, and compares the relative success and failure of the businesses that competed to supply what continued to be the single most important item in the optical armoury, the large naval rangefinder.

The Industry and the War Office's Influence

In 1907 the British Army was still far from a large-scale user of optical munitions and much of what it employed was already acknowledged within the service as unsatisfactory and obsolescent. Little had been done to rectify the shortcomings demonstrated in the Boer War and the Army's spending on optics since then had averaged only £1,700 a year.[1] That changed after 1908 when decisions were made to adopt new instruments, causing spending to increase and generating new business for optical manufacturers. It has been previously suggested that not only did the War Office do little to support the domestic optical industry up to the outbreak of war in 1914, but through a combination of favouring foreign makers and distributing orders piecemeal amongst British companies it actually discouraged the home industry from becoming involved in military contracting.[2] The pattern of ordering, however, does not bear out this view, showing rather that War Office business was concentrated on a few British firms who thus became progressively more experienced in optical munitions production. That is not to say that the British Army was a prolific spender or that, unlike the Admiralty, it deliberately encouraged the home industry, but nevertheless military orders were placed at an increasing pace after 1910. The interaction of the War Office with the optical manufacturing community can best be shown by examining the process of selection and purchase of three key types of optical munitions – the single-observer rangefinder, the panoramic artillery gun sight (the 'dial sight') and the prismatic binocular – all of which began to be seen as essential equipment for the efficient prosecution of warfare. Together, these accounted for over 90 per cent of the £175,000 that was budgeted for optical purchases, the records of the Army Contracts Department from April 1907 to 31 March 1914 showing that approximately £49,000 was spent on rangefinders, £66,000 on dial sights and £45,000 on binoculars.[3]

The Rangefinder

In 1907, the British Army still lacked a satisfactory infantry rangefinder despite the numerous trials and evaluations of the Marindin design described in the previous chapter. Only in January of that year was it finally considered necessary 'that an infantry rangefinder should be immediately supplied' and the following month the Marindin was formally approved for service.[4] The quantities needed were, however, uncertain, because the scale of issue for rangefinders was currently being reviewed. If the Marindin simply replaced the earlier Mekometer, then only 300 would be bought, but if a wider proposed issue were adopted then the total for the entire infantry of the front-line forces would be 1,040.[5] Budgeting for either quantity in the Army's financial estimates was essential but somewhat problematic as at no point during its evaluation process had any detailed costs been requested by the War Office.

Captain Marindin, the inventor, had his trial instruments made by Adam Hilger & Co., but there had been no formal liaison between the firm and the Army and the only price mentioned had been Marindin's informal estimate of £35 if what he vaguely described as 'very large' numbers were ordered. The Master General of the Ordnance was being pressed to organize the rangefinder's early introduction even before its formal adoption, and he proposed to supply the troops with 300 as Mekometer replacements during the next two years. He thought Marindin's estimate unrealistic as the trial instruments had cost at least £85 each, and he reckoned the likely production cost would be around £50, probably using as his only available yardstick the current price of the Barr & Stroud infantry rangefinder which that company was then offering for sale. Having decided on the need to buy 300 rangefinders at a likely cost of £15,000, the Master General could only afford to allocate £5,000 for them in the Annual Estimates for the fiscal year 1907–8. Despite that difficulty, he was optimistic enough to say that, somehow, 'steps will now be taken as to ... obtaining a supply' of the rest of what was needed as a minimum for re-equipping the infantry.

His optimism was somewhat misplaced. Obtaining any supply was complicated because Marindin and the War Office were by then in dispute about the question of financial reward.[6] When asked in 1902 about his terms for making the rangefinder available to the Crown he had valued it at £25,000, but no further discussions took place until Marindin learned in early March 1907 of the decision to adopt his rangefinder and immediately resurrected the matter. Pending an answer from the War Office, he then took the highly unusual step of refusing to hand over its detailed drawings to the Chief Inspector of Optical Stores at Woolwich Arsenal, thus stopping the War Office from drawing up the specification which it needed in order to request tenders for manufacture. If his most likely reason for doing so was that he wished to prevent the War Office exercising what amounted to its right of sequestration over patented inventions without first agreeing details of payment for their use, then the impasse he created was both short-lived and unsuccessful. It was broken by the Secretary of State for War who reminded Marindin that he was 'withholding the information necessary for the manufacture of the instrument for His Majesty's Service' which, in practical terms, amounted to an order to surrender the details immediately. As a serving officer, he had little alternative but to acquiesce and hand over what was needed, having to trust to fortune about his eventual reward.

A request for tenders was issued by July 1907, but it was another four months before any contract was placed. In November Adam Hilger & Co., who had made all the rangefinders so far, was given an order, not for the hundred which had been budgeted for, but for just sixteen instruments for troop trials.[7] That was because the War Office had decided that Marindin's claim for reward could only be judged after seeing how well the rangefinder performed when issued to 'ordinary' infantry units. There are two possible inferences from this seeming retreat from the earlier enthusiasm for the instrument. Either the rangefinder had lost its allure to the infantry, or there had been a shift by the War Office to the notion of the linking of its monetary value to its demonstrated utility in general service. Whatever the underlying reason, the War Office was reluctant to commit itself to further purchases until a firm offer had been made to Marindin. That was eventually made in June 1908, but still only fifty were ordered despite the supposedly urgent need to have it in service and funds for a hundred already being available. The continued unwillingness may have been because Marindin rejected the offer as inadequate and then appealed to the Treasury for the figure which he had proposed in 1902. That process took a further year and, perhaps unsurprisingly, failed to give him the £25,000 requested. Instead, the Treasury allocated him a royalty of just 15 per cent on each rangefinder accepted for service, plus his earlier expenses.

Even when the question of Marindin's settlement was finally out of the way, his rangefinder was never ordered on a scale large enough to equip the entire front-line infantry force. The number eventually ordered up to April 1914 was just 337 at a total value of £22,305, which was far less than the expanded scale of issue called for. Those issued proved less than satisfactory in service and its sole maker, Adam Hilger & Co., was unable either to eradicate the problems or to deliver those ordered on schedule. The firm never managed to produce it at a rate greater than two per week, and, adding to the War Office's procurement difficulties, no other maker was ever inducted into its production.[8] The only other firm to tender and supply sample instruments was Thomas Cooke & Sons Ltd in 1908, but presumably either the quoted price or the functioning of samples was unsatisfactory, as the company never received any production contracts.[9] The War Office's adoption of the Marindin rangefinder provided few benefits for the industry, either through large orders or any kind of spin-off that might have opened up new avenues for its maker. Adam Hilger's problems that restricted the company benefiting from it will be described later in this chapter, but the rest of the optical industry was doubtless deterred from competing to produce it principally because the technical and logistical difficulties in setting-up outweighed the guarantee of reward. Without the assurance of substantial orders no business was willing to tackle a complex manufacturing problem that was outside its prior experience.

Eventually, in 1912, the War Office ordered a small quantity of infantry rangefinders from Barr & Stroud and the next year followed what had by then become the example of almost every other European army and began to buy them in bulk. Orders in 1912 totalled £4,313, rose in 1913 to £9,724 and in the first seven months of 1914 leapt to £54,000, almost two and a half times the money spent on the Marindin in the previous six years.[10] The War Office also spent £13,055 on German Zeiss rangefinders for experimental issue to the field artillery in 1911 and 1913, which might be construed as a lack of confidence in the domestic industry. This is doubtful; a far more likely explanation is that those trial purchases demonstrated the artillery's continuing inability to decide on what design of rangefinder it actually wanted, rather than discrimination against domestic models. Having vacillated since the end of the Boer War about what was its ideal type, by 1911 the artillery branch had eventually progressed to the point where it had accepted that the single-observer pattern was acceptable. Even by the summer of 1914 with the prospect of war in Europe seeming increasingly likely, the artillery was still deliberating over what was required.

Although its rangefinder procurement process was long drawn out and failed to produce early or substantial benefits to the optical munitions industry, the War Office's efforts to standardize satisfactory types of artillery sights and prismatic observation binoculars went far more successfully. Yet, those very processes have led to the idea that, far from the War Office stimulating the capacity for optical munitions production before the First World War, the way it went about doing business led to the active discouragement of British firms operating in this area.[11] According to this understanding, the War Office did so by purchasing many of its new instruments from Germany, and by spreading small contracts for the remainder across different domestic firms in a misplaced tactic to stimulate competition. This, supposedly, had the unfortunate opposite effect of forcing up prices and ultimately discouraging their mass production in Britain. Although the War Office was frequently less than adroit in the way it did business, its practices in the five years before the war began were actually improving and, in this case at least, it really does not deserve such censure.

The War Office and the Artillery's 'Dial Sight'

This was one clear instance when improvements in optical technology combined with a newly perceived need resulted in the creation of an entirely novel instrument which became a key component in the effective deployment of land artillery. The Russo-Japanese war demonstrated that artillery could often best be used from positions which were out of sight of the enemy, thus removing the possibility of retaliatory fire. Furthermore, unlike in South Africa the topography of Southern Manchuria meant that field guns often had to be placed in positions where it was impossible for their sights to be laid on the intended target.[12] Engaging such obscured targets by 'indirect firing' was not new to gunners, and was accomplished by aiming-off from a proxy target whose angular displacement from the actual one had been measured, so that the gun's sight could be set to an appropriate deflection in order to point the barrel in the correct direction.[13] The displacement angle was measured using surveying techniques, the sighting done with an instrument using an aiming telescope fixed to a large horizontal, precisely divided dial – hence the name – which was bulky, awkward to use and relatively fragile. In 1904, the Berlin company, Optische Anstaldt C. P. Goerz, introduced a new, radically different type of dial sight using a complex optical system.[14] This miniaturized 'panoramic' dial sight was a compact prismatic aiming telescope that also functioned as a periscope which could traverse through a full circle whilst maintaining a magnified normal – i.e. neither inverted nor reversed – image for the observer. It provided for both indirect and direct aiming, which benefited its operator who could now remain protected behind the gun's shield. Its introduction was possible because Goerz had devised a novel prism system which, unlike earlier attempts, maintained the normal image as it was traversed. The optical system was complicated and, unsurprisingly, closely protected by international patents.

The Goerz panoramic sight was quickly noticed by the artillery, and the War Office organized tests in Britain soon after the device became available in 1904. As with rangefinders, trials went on for several years largely because the artillery wanted to make substantial modifications to the original Goerz design in order for it to be better adapted to British service requirements. Eventually a design was finalized with Goerz and the now greatly modified instrument approved for service in 1909 and ordered as the 'Dial Sight No. 7' (see Figure 3.1).

Figure 3.1: 'Dial Sight No. 7'. The battered instrument depicted here is a 'battlefield survivor' made by R. & J. Beck Ltd in 1917. Approximately 12 in. long, the sight was mounted vertically on the gun. Author's collection.

The sight was seen as highly important for the effective deployment of field artillery, especially as the War Office had begun to take delivery of new, improved patterns of highly mobile guns. Far more effort went into standardizing the sight's design and organizing its procurement than was the case with a new artillery rangefinder. It was also taken into use more quickly and successfully than the infantry rangefinder, and on a larger scale. Far from being handled in a manner that disadvantaged the domestic industry, it represented something of a success story in the procurement and manufacture of optical munitions and the stimulation of the industry.

Even though the sight had been greatly redesigned to suit the War Office, it was still protected by the international patents held by Goerz. The War Office was unwilling to rely on a foreign source for supplies and so negotiated a licensing agreement to allow it to be made in Britain. This was essential, because in the five years since the panoramic sight's arrival, no other optical company had been able to devise a similar instrument without infringing the Goerz patents. This should not be taken as indicative of a weak British optical industry; not even Zeiss in Germany, by then generally acknowledged as the world's foremost innovator in complex optical systems, had been able to produce a viable competitor. The arrangement brokered with Goerz was actually decidedly favourable to British interests. It required the purchase of some instruments directly from their factory and the payment of royalties for those made in Britain. Only 30 per cent of the orders were to go directly to Goerz, a relatively small proportion for a product on which the company held a seemingly unassailable monopoly. That the balance of the business was to come to British industry might have been the result of hard bargaining by the War Office, but it is more likely that another factor altogether influenced the willingness of Goerz to forego a greater degree of involvement.

By the time the War Office placed its first contracts for the sight, the Companies (Consolidation) Act of 1907 and the new Patents Act of 1907 had come into force. These were essentially protectionist pieces of legislation which stood to have an adverse effect on foreign companies trading in Britain.[15] The new Companies Act required them to disclose full details of their parent company's financial affairs, wherever they were conducted. Under the new patents legislation, if a British patent held by a foreign patentee was not being 'worked' on a commercial basis in Britain, then the patentee was obliged to grant a licence to any 'interested person' who wished to take it up. The response of Goerz, and at least one other German optical manufacturer, was to form British companies. In 1908 Goerz had set up a London subsidiary, the C. P. Goerz Optical Works Ltd, with an initial share capital of £5,000 that was subsequently increased to £10,000. That step let the company maintain its overall business confidentiality and opened the way to keeping a close hold on the design through the licensing arrangement with the War Office.

Those steps may have led the War Office subsequently to consider Goerz as a British supplier. The War Office Contracts Department identified all foreign purchases in its yearly reports, but only two of Goerz's four contracts were so described, in the financial years 1910–11 and 1911–12; those in the next two years were listed as domestic ones. Despite that, it is not clear if Goerz actually made, or even assembled, instruments in Britain. The only published description of the firm's British activities suggests that, unlike the larger firm of Zeiss, which did set up a manufacturing business in addition to its import and marketing structure, it remained no more than a marketing company – 'einer Vertriebgesellschaft'. However, its share capital of £10,000 was the same that Zeiss employed in its combined manufacturing and distribution activities, raising the question of why such a large amount was needed by a smaller business operating on a lesser scale. In the absence of other evidence any answer must be conjectural, but as far as the Army's purchasing department was concerned, by April 1912 the company was clearly being treated as a British supplier.

Between April 1909 and April 1914 the dial sight accounted for more expenditure than any other single-optical store purchased by the Army Contracts Department – £65,698, or 37 per cent of the total expenditure on optical munitions, with 1,662 sights eventually being contracted from six makers.[16] Sufficient numbers were ordered to match the gradual introduction of the new guns meant to be the principal recipients of the sight. Between 1904 and July 1914, approximately 1,650 new guns were ordered for the British and Indian Armies, and although orders for dial sights belatedly began in 1908, they did eventually match gun deliveries.[17]

Table 3.1: Dial sight contracts, 1909–14.[18]

Maker	1909–10	1910–11	1911–12	1912–13	1913–14	Total	%
Cooke/Vickers	15	nil	nil	63	nil	78	4.7
Barr & Stroud	55	nil	25	nil	nil	80	4.8
Beck	15	nil	125	130	222	492	29.6
Goerz	nil	168	100	100	124	492	29.6
Ross	nil	20	100	196	204	520	31.3
Total	85	188	350	489	550	1,662	100

The 70 per cent of dial sight orders placed in Britain were spread between six companies, seemingly supporting the assertion that War Office contracts were spread too thinly to be attractive to makers. Barr & Stroud, Beck, Cooke's of York, Goerz, Ross and Vickers all received orders (see Table 3.1). However, Vickers actually had no optical capability, and its share of the contract would have been made by Cooke's which had business connections with Vickers.[19] Some 85 per cent of the British orders were divided between just two firms, in roughly equal proportions. To what extent this was through a misplaced desire to stimulate competition is open to debate, but the assertion that it drove up prices is certainly not justified. The original Goerz order in 1910 was at £40.05 per instrument, a figure repeated in 1911 and 1912, and the final Goerz contract in 1913 was lower at £38.00. Goerz was already making a similar, though not identical, sight in large numbers for the German forces, and, allowing for the resulting economies of scale and lower German wage costs, it might be expected that the British firms' prices would have been significantly higher. Beck's were more, but only by 5 per cent: £42.50 in 1911, £42.20 in 1912 and £40.00 in 1913. The prices from Ross were actually cheaper – £35.00 in 1910 and 1911, £37.50 in 1912 and £37.75 in 1913. Although one of the Contract Department's main responsibilities was to ensure that prices charged were reasonable, it also had to ensure deliveries were made at rates appropriate to service requirements, and the division between Beck and Ross may actually have been necessary to obtain the numbers required to match gun deliveries.

The idea that the War Office distributed orders in small numbers primarily to stimulate competition is open to question. The size of contracts was governed both by the funds available and the timetable of need. Even under pressure to provide an effective infantry rangefinder 'immediately' at a cost exceeding £15,000, the War Office had only been able to budget £5,000 in each of the current and following financial years because that was the limit of funds available.[20] Spending on munitions contracts in 1907 was the lowest for twelve years and orders for weapons for the Army until the end of 1910–11 continued to be lower than even before the Boer War.[21] The need for dial sights was geared to the delivery of new artillery weapons whose rates of production were initially slow, so with limited budgets it made no sense to contemplate ordering in advance the full outfit of sights for the whole gun programme. Fiscal prudence, or necessity, rather than misguidance would better account for the absence of larger-scale dial sight orders. A single large contract might have resulted in economies of scale which could have reduced production costs, but the financial conditions to place such an order simply did not exist, even if any single contractor had been able or willing to take on the work. Although ordnance spending was low, the overall level of commercial trade was good, so that manufacturers were, for once, not desperately seeking new business.[22] If the War Office had offered one very big order for the dial sight requirement it may well have found itself having to persuade or even tempt instrument makers to take the contract when they were not particularly eager to bid for government work. Such a situation would hardly have driven prices down.

Both the War Office and the optical manufacturers seem to have emerged creditably from the dial sight procurement programme. The delays in ordering were because of the War Office's requests for changes in the design. Once that was settled the state not only secured a broadly favourable licensing agreement with the German patent holder, but secured deliveries at a rate appropriate to the arrival into service of the guns with which it was meant to be used. The optical munitions community was expanded and enhanced by its familiarization with a novel and highly complex instrument requiring the highest standards in optical and mechanical engineering. The accrued, if latent, benefit to both sides was the provision of a manufacturing base which, as will be seen, could be substantially expanded in the event of war.

Many of the same factors also applied to the prism binocular, the third category of optical munitions ordered in substantial numbers up to 1914.

The Prism Binocular

The place of the prismatic binocular in the inventory of optical munitions was very much different to either the rangefinder or the dial sight. First, it was not something which only had application within the armed forces; it did indeed have a substantial civil market and even by 1907 was being produced commercially by a number of British optical instrument makers to whom the War Office could turn for supplies. The Army had found itself lacking efficient patterns in its most recent war and by 1907 had become disposed to equip itself with the prismatic type which offered the benefits of higher magnification and a wider field of view over the simpler galilean type that had been the earlier standard pattern. Binoculars for military use were essentially the same sort as those already freely available although, like all War Office purchases, they were subject to rigorous quality control and inspection procedures. Their procurement ought not, in theory at least, to have been difficult to manage. Despite all this, the prismatic binocular, like the dial sight, has been cited as another instance of the War Office's purchases of German instruments failing to support a domestic industry which was itself a fragmented body of family firms of limited abilities.[23] This was far from the case and there were actually many similarities with the programme for acquiring dial sights.

Between April 1908 and the end of March 1914, the War Office ordered approximately 11,700 prismatic binoculars, which was enough to equip adequately the peacetime regular army on the official scale of one for every twenty officers and men.[24] Their total value was some £51,000 of which only £7,500, less than 15 per cent, was directly spent abroad, and all of that during the first year of purchasing.[25] Binocular acquisitions began in 1908 when 1,500 were ordered from the German firm of Carl Zeiss through its sales office in London. The reason for this initial foreign order was that although prism binoculars were being made in Britain, only those produced by that German company at that time had the design element which provided the observer with an enhanced sense of three-dimensional vision at far distances. That was achieved by arranging the objective lenses so that they were much wider apart than the human eyes' separation, a design feature protected by a patent which was due to run out in 1908. After that time, British companies would be free to incorporate the feature in their production.

The following year, 1,000 of a similar pattern, designated 'Binocular No. 2', were indeed ordered from British companies which had set themselves up to take advantage of the patent's expiry (see Figure 3.2).

Figure 3.2: 'Binocular No. 2'. Introduced in 1908, it was manufactured even beyond the end of the First World War. The one illustrated was manufactured by Ross in 1917 for an Admiralty contract. Author's collection.

Purchases subsequently increased year by year, peaking in the financial year 1913–14 when 3,031 were bought. As with dial sights, however, the War Office was prepared to class a foreign owned company as British supplier. In 1909, driven by the same considerations as Goerz with its dial sight, Zeiss set up a British subsidiary manufacturing company at Mill Hill, London, which produced binoculars whose components were, apparently, sent from Jena for finishing and assembly.[26] This new business, Carl Zeiss (London) Ltd, received approximately £5,600 worth of orders between 1911 and 1914 which, as was the case with Goerz, the War Office treated as domestic business in the annual reports of the Contracts Directorate.[27] The Mill Hill Zeiss works was jointly managed by German and English staff, but its workforce was predominantly English and its binoculars were marked as though made entirely in England. Even if all the business placed with Zeiss were counted as foreign, then the approximate total of £13,100 was still less than 26 per cent of all prism binocular orders, which hardly made Germany the chief supplier, or the War Office reliant on foreign instruments. The greatest proportion of binoculars was bought from the Ross Optical Co. of London, which supplied almost 5,000 of them. Watson and Son, also of London, produced over 3,000 and the third unarguably British firm, Aitchison & Co., made some 600. The London arm of Zeiss delivered almost exactly the same number as did its German parent in 1908–9.

The War Office may have deserved censure because of the way it handled its rangefinder needs, but its other optical munitions business was geared to a relatively small army which was only a quarter the size of Germany's; it could never have offered the optical makers orders on the scale that the German industry received from its own War Ministry.[28] Its hesitant selection procedures may have retarded the placing of orders, but nevertheless it was still a substantial client of the emergent optical munitions community. In the six fiscal years preceding the start of the First World War the War Office spent over £179,000, roughly £17.3 million in 2011 values, buying more assorted items of optical munitions, generally ordering as many instruments as the Army actually needed. It certainly did not deliberately place obstacles in the way of the domestic industry, nor was it so badly served by British contractors that it had to depend on German imports. Its suppliers were a core group of specialist British makers which had emerged from within the general optical instrument making community and, so far as making these devices was concerned, were separate from it. Far from scattering small contracts across the greater optical trade before 1914, the War Office actually concentrated its orders on a small number of firms who accordingly became familiar with producing instruments to the exacting though sometimes frustrating standards demanded by the Army. Nevertheless, although the scale of War Office spending increased greatly between 1907 and 1914, it was still very much less than the Admiralty's, which provided a larger and very different market for optical munitions.

The Admiralty's Relationship with the Industry

The Admiralty continued to be a much larger customer for optical munitions than the War Office up to the outbreak of war in 1914, and rangefinders dominated its orders. The Royal Navy's demand for them was substantial and of very considerable value, usually outweighing that for other optics. Rangefinder requirements were linked to a substantial shipbuilding programme intended to maintain a margin of superiority over other navies, and which is usually associated with battleships and battlecruisers.[29] The lesser vessels such as cruisers and destroyers that were also built added to the scale and variety of demand for optical munitions. Although capital ships required the largest and most sophisticated types, cruisers were to be provided with outfits of optical instruments which, only a decade earlier, would have been seen as lavish even on the biggest warships. In addition, from 1907 the Admiralty began to ask for rangefinders that could be used on the smaller vessels such as destroyers, which carried weapons of lesser range and lacked the space to mount the 9-ft base models which were becoming standard on larger vessels.[30]

The Royal Navy not only purchased rangefinders. It bought many other optical instruments, particularly for its larger ships. Telescopic sights and observation telescopes of increasingly sophisticated design were needed for the gun turrets of capital ships, as well as simpler sighting telescopes for their secondary armament.[31] Similar sights were also required for the cruisers and destroyers built during this period. The massive 're-sighting' programme of 1905–7 described in the previous chapter had provided only for ships in commission or about to complete. The construction of new capital ships with heavier, longer ranging armaments coupled with increasing attention to gunnery meant that by mid-1914 the Admiralty was steadily requiring new and varied types of optical munitions. Orders for these, although on a much smaller scale than in 1905–7, continued to be placed amongst the community of contractors with whom the Admiralty had already built up working relationships. Barr & Stroud supplied rangefinders, Ottway and Ross made gun sighting telescopes and Ottway, Ross and Thomas Cooke's of York – which supplied the instruments to Vickers – produced observation and sighting periscopes for surface vessels.[32] Submarine periscopes were almost entirely made by Sir Howard Grubb & Co. of Dublin, who made them under contract to Vickers at a time when they had a virtual monopoly of submarine building for the Royal Navy.

Greater demand came not only from a growing modern fleet after 1907, it was also driven by efforts to improve the probability of hitting distant moving targets as fighting ranges of 10,000 yds and more were being increasingly envisaged.[33] Shooting at such distances necessitated some means to predict where the moving target would be at the end of a projectile's flight time, which at 10,000 yds was more than 15 seconds.[34] The concept of 'fire control' – a systemized means to direct a ship's guns against a moving and distant target – became gradually accepted as essential, and increasingly complex electrical and mechanical systems were developed using the optical rangefinder as the primary means for generating the required data. The development of those systems has been examined in considerable detail elsewhere, but little attention has been given to the optical instrumentation involved. Attention has concentrated on the evolution of the mechanical computing aspects of such schemes, and in particular the one devised by the civilian inventor Arthur Pollen, which competed unsuccessfully against another one devised by a serving officer, Captain F. C. Dreyer.[35]

Fire-control needs drove the demand for more accurate rangefinders, and the Admiralty began to call for instruments to measure accurately at distances considerably further than the longest range at which shooting was expected to start, in order to collect range and bearing data from which the target's future position could be predicted. By October 1907 the Admiralty wanted to measure ranges of 15,000 yds with an accuracy of 1 per cent in order to open fire accurately when the distance eventually closed to 10,000 yds.[36] This standard was beyond the capability of the instruments then in service, although rangefinders to do a similar task had already been mooted by the Imperial Russian Navy in 1906.[37] The growing stress on greater accuracy and longer ranges helped to stimulate the development of larger and more complex instruments as part of a system of gunnery, emphasizing that large naval rangefinders could no longer be seen as isolated from the rest of a ship's armament. It was this evolutionary state of fire-control instrumentation that introduced Thomas Cooke & Sons Ltd of York as potential commercial rivals to Barr & Stroud, and a comparison of their progress during this period offers some insight into the variety of technological and social forces acting on and within optical munitions contractors.

Thomas Cooke & Sons Ltd: A Competitor for Barr & Stroud

Cooke's of York was no stranger to optical munitions, although its previous products had been relatively simple in optical design and construction. The firm's chief designer, H. D. Taylor (1861–1943) seems first to have been directed towards optical munitions during the Boer War when he designed an optical sight to improve the accuracy of shooting at long ranges. He was granted a patent in connection with rangefinders in 1903, soon afterwards obtaining two more, relating to a novel layout and the use of rotating prisms intended to produce a high level of robustness.[38] Between 1904 and 1906 Cooke's built five different experimental models to his designs, culminating in the unsuccessful submission of a 10-ft instrument for Admiralty trials against Barr & Stroud's latest 9-ft model.[39] During this period, Cooke's had come into contact with Arthur Pollen through making the optics for his own abortive design for a two-observer rangefinder in 1905,[40] an experience which seems to have encouraged the firm to delve deeper into rangefinder design as the connection between it and Pollen grew stronger. In 1907, Taylor began to refine his earlier efforts and between then and 1911 was granted seven more patents covering a range of increasingly sophisticated designs.[41] The two experimental models resulting from them were superseded by a radically new design in 1912 that was meant to form an integral part of Pollen's fire-control system and to be sold as a component of it.[42]

Pollen's involvement with Thomas Cooke & Sons Ltd came firstly through his need for the high precision mechanical engineering of the sort that Cooke's employed in survey instruments and astronomical telescope clock controls in order to produce the high precision cams and intricate gearing used in the mechanical analogue computer he was developing.[43] In 1908 Pollen became a shareholder and director of the firm,[44] and so created for himself the opportunity of also using Cooke's optical skills to develop a complete fire-control system including a sophisticated rangefinder which might be sold as a patented package. This constituted potentially serious competition for Barr & Stroud because of Cooke's considerably greater optical design capabilities. H. D. Taylor was an internationally recognized expert in the design of telescope optics and camera lenses and, unlike anyone at Barr & Stroud, he was well able to compute increasingly complex lens and prism systems which Cooke's by then were capable of making entirely by themselves.[45] Barr & Stroud saw Cooke's involvement with rangefinder design, either on its own or through Pollen's marketing arm, the Argo Co., as giving the Admiralty the prospect of an alternative supplier and possibly ending the monopoly previously guaranteed by being the only British maker. Barr & Stroud was taking Pollen seriously as early as 1908, when its general manager, Harold Jackson, instructed his resident engineer at the Royal Dockyard in Portsmouth to find out all he could about Pollen's activities and plans, writing enigmatically in the December that 'we understand ... he is on with something'.[46] The 'something' was not then clear, but by March 1911 Jackson knew a great deal more and believed that Cooke's and Argo in combination would 'in all probability shortly be serious competitors'.[47] The Admiralty, although totally committed to buying British-made instruments, was by no means contracted to one domestic supplier in perpetuity.[48]

Despite Barr & Stroud's concerns over the possible competition of the Cooke–Pollen rangefinder, such fears were probably misplaced because the Admiralty had a number of forces acting on it to shape its policies concerning rangefinders. These included not just technological issues but also cultural and political ones that had sometimes subtle, but sometimes very direct, influences on its decisions.

One advantage that Barr & Stroud undoubtedly had over Cooke's and Pollen was the existence of beneficial contacts within the Navy itself. One of those was Professor J. B. Henderson, who had earlier worked with William Stroud in Leeds and then with the firm in Glasgow as head of its scientific research department. Henderson was appointed Professor of Applied Mechanics at the Royal Naval College at Greenwich in 1905.[49] He subsequently corresponded regularly with his old employers, and in October 1907 wrote privately to Stroud to advise him of the influence Pollen's ideas were having on naval gunnery, and in particular of the problem of hitting moving targets at very long ranges. Henderson not only told Stroud that a rangefinder of much greater accuracy would soon be called for, but also directed him diplomatically towards the idea of becoming involved in fire-control instrumentation by saying 'Pollen is a fairly skilful mechanical inventor, but he is not a scientist and cannot tackle the problem'.[50] Events would prove Henderson wrong about that, but his condescending opinion of Pollen's abilities did persuade Barr & Stroud to take up the idea. Stroud had previously worked on the design of the firm's electro-mechanical 'Range-and-Order Indicators' which transmitted range and other gunnery information to individual gun mountings and were the rudimentary precursors of what Henderson was now discussing, so he would have appreciated the amount of work and the complexity of the problems likely to be involved. That may have persuaded him that were the idea to be taken up successfully, it would require more expertise than the firm had available, leading the company to begin a collaboration with the Dutch artillerist and engineer Admiral W. Mouton that proceeded until temporarily interrupted by the outbreak of war.[51]

As with the earlier Range and Order indicators, Barr & Stroud saw fire-control instrumentation as an extension of its activities rather than diversification, although to what extent it saw the Admiralty as its main client for such new products is uncertain. Given the Admiralty's long-running dealings with Pollen and Dreyer which were by then no secret, the firm may have seen the new product as wholly export-oriented from the outset. A delegation from the Imperial Japanese Navy examined one of the earliest versions of what was called the 'predictor mechanism' in March 1912 at a time when they were ordering large amounts of rangefinders and Range and Order instruments.[52] Progress on the predictor was slow, however, and the developed prototype had still to be finished when the war began. It was then put into abeyance for the war's duration and, perhaps literally, dusted off in 1919 for its subsequent completion.

Whilst Pollen's relationship with the Admiralty was frequently less than harmonious, leading to distrust and even hostility, Barr & Stroud retained a significant degree of confidence from the Navy despite potential conflicts of interest over the amount and nature of foreign trade the firm carried out.[53] In 1908, the Director of Naval Construction asked for an assurance that foreign officers visiting the factory would not be able to see any 'confidential work' being done for the Royal Navy, to which Jackson had to reply diplomatically that there was actually nothing being supplied to the Admiralty that had not already been sold abroad. He refrained from adding that much of it had been sold to foreign navies even before the Royal Navy had adopted it. Despite numerous earlier offers to keep designs secret, he said, the Admiralty had never taken them up and the firm had repeatedly been told it was free to submit them to foreign governments, which it had then done.[54] Jackson was obliged to point out that in consequence foreign trade had become so important that the company could no longer afford to disregard it. Irrespective of whether he was mollified or chastened, the Director let the matter drop, but four years later, on a different tack, he asked for details of what foreign navies were ordering and whatever else they were asking about. Jackson responded that as he had no specific instructions from any overseas client to observe confidentiality he considered the firm was 'quite at liberty' to tell the Director whatever he wanted to know.[55] By that time, much of the Admiralty's work was sufficiently different to foreign contracts that a special department had been set up to handle it and soon after telling the Director exactly what every foreign power had ordered recently, Jackson asked if he could allow trainee rangefinder technicians from the Imperial Japanese Navy into the rest of the factory as orders from Japan were 'by no means inconsiderable'.[56] The Director was quite happy to permit this, evidence that both parties were tacitly recognizing the symbiotic relationship that had developed between them, something that Cooke's lacked and which they were never able to cultivate, very much to their detriment.

The intricacy of Cooke's association with Pollen and the Argo Co. may not have been clearly understood by Barr & Stroud, but the construction and significance of Taylor's 1912 rangefinder design mentioned earlier most certainly was.[57] Jackson described its principal features to the firm's Austrian agent in July 1912, detailing its novel optical design and gyro-stabilized data-transmitting mounting, both of which he had to concede Barr & Stroud had nothing to compete with. Putting a brave face on it, he observed the rangefinder was 'very complicated and ... very costly' but had to concede that its unusual optical system provided 'extra brightness' that made it more useful in the bad lighting conditions typified in the North Sea. He also noted that the gyro-stabilized mounting let the operator take readings more quickly and certainly than either the pedestal or turret mountings provided by Barr & Stroud. Despite this recognition, the firm did nothing to embark on a similar system for its own rangefinders.

Although Barr & Stroud had serious concerns about the threat from the Cooke–Pollen rangefinder, it stood little chance of being adopted by the Admiralty. There were several reasons. First, it was seen as an integral part of Pollen's fire-control system, which he was struggling with increasing difficulty to persuade the Royal Navy to accept. By 1912 it was inclined to prefer a less complex, and also less ambitious, system designed by a serving officer, a circumstance not unlike the War Office's situation with the Mekometer in 1889. By 1914, when the trials finally ended, the Pollen system was rejected, and with that went Cooke's chief hope of selling the complex rangefinder to the Admiralty. Adding to the difficulties of association with Pollen in selling it to the Royal Navy, Taylor's design had constraints that would have made it highly unlikely that the Admiralty would have considered it as a replacement for the existing Barr & Stroud patterns.

Taylor's rangefinder provided a brighter image of higher contrast than the Barr & Stroud models, which enhanced its use in adverse lighting conditions.[58] This had been achieved through Taylor's ability entirely to redesign the telescope portion of the rangefinder to benefit from the properties of new advanced optical glasses being made by Schott & Genossen of Jena in Germany, which permitted substantial improvements in the performance of telescope lens sys-tems.[59] Barr & Stroud had never used these glasses, partly because the firm had no designer of sufficient ability to compute systems around them, and partly because by using the older flint and crown glasses readily available in Britain it was possible to make optics that were still generally satisfactory for use in a range-finder. Taylor had used the new Jena glasses in his designs for camera lenses and astro-telescopes almost as soon as they became available, had suggested modifi-cations in their formulation to Otto Schott, their inventor, and become firmly wedded to their employment wherever possible.[60] Even by 1914, only a few of the increasingly wide range of Schott's sophisticated types of optical glass were yet being made in Britain. Their formulae were complex and their production processes far from easy to master which, combined with the still-limited British demand, had hardly encouraged their production by the country's only optical glass maker, Chance Brothers of Birmingham. Taylor's use of those glasses in the rangefinder would have posed problems in selling it to the Admiralty in view of its insistence on domestically made glass for all its optical instruments.

That policy was rooted in the need to be independent of foreign suppliers in time of war. Optical glass manufacture was done in Britain, but since the 1890s the German and French industries had become larger and more adept at manufacturing a wide range of specialized glasses, many of which were unavail-able elsewhere. In 1910, fearing that relying on imported optical glass would lead to severe problems if supplies were interdicted by an enemy, the Admiralty had begun to stipulate that British optical glass should be used wherever possible.[61] Consultations were encouraged between the instrument makers and Chance Brothers to assure supplies of both the established and new formulations. Chance already made a wide range of optical glasses, but attempts to get them to produce domestic alternatives to the new ones met with only limited success. The firm saw optical glass as an unprofitable aspect of its business, it lacked both the tech-nical staff and facilities to make rapid headway in catching up lost ground, and was unwilling to invest heavily in the development of material for which it saw little profitable outlet.[62] This greatly restricted the range of more sophisticated glasses available in Britain but represented little difficulty for Barr & Stroud who had already given a categorical assurance in 1911 that they were independent of imported material. Because Cooke's had never been an Admiralty supplier, Tay-lor may not have been fully aware of the problems he had created in producing a rangefinder that depended on what were now in effect proscribed raw materi-als for its much of its optical superiority. In the end, though, the Cooke–Pollen rangefinder was rejected for an altogether different reason.

Its failure came not because it was part of a larger rejected system, had unfa-vourable associations with the now out-of-favour Pollen or used unacceptable materials, but simply because in its 1914 trials it consistently failed to read ranges accurately. Its obituary notice pronounced by those testing read, with commendable restraint, that 'It is a beautiful instrument but it has one serious defect, namely that it will not measure distances'.[63] Taylor's sophisticated range-finder failed through mechanical difficulties that, as with Archibald Barr and William Stroud's earliest models, might well have been remedied by revision and modification. Unfortunately for the makers the declaration of war only three months later ended its chances of success and Cooke's were, to all intents and purposes, kept out of the business of making large naval rangefinders.

Cooke's failure to break into the rangefinder market was not because the firm lacked optical expertise, but because it was coming, at an inopportune moment, late into a field where the Barr & Stroud models had already established a considerable technological momentum. For the Taylor-designed instrument to have displaced them would have required first the clear demonstration of its superiority and second the institutional willingness to accept an alternative to the existing perception of rangefinding instruments which had become wholly identified with the Barr & Stroud models in service.[64] It was perhaps ironic that Barr & Stroud, a business far less able in optical design, should enjoy a conspicuously greater degree of success selling complex optical munitions.

Barr & Stroud's Growth

Barr & Stroud enjoyed considerable success between 1907 and 1914, both in sales and profits, and its domination of the rangefinder market. The firm generated a substantial income during those seven years, with rangefinder sales totalling £806,000 and earnings after tax £240,381, a healthy net margin of 29.8 per cent.[65] The greater proportion of those earnings actually came from foreign business. The company's records show the distribution of orders, and provide a picture of the extent of the company's dependence on foreign armies and navies (see Table 3.2).

Table 3.2: Barr & Stroud, total orders received from the British Admiralty, War Office and all foreign clients, January 1907–July 1914 (amounts shown in sterling).[66]

Financial Year	Admiralty £	War Office £	Foreign £	Total £
1907	60,382	nil	23,490	83,872
1908	31,112	155	53,464	84,731
1909	23,521	438	84,746	108,705
1910	42,904	2,517	64,178	109,599
1911	32,172	5,678	50,241	88,091
1912	36,148	4,313	160,768	201,229
1913	72,966	9,724	185,330	268,020
1914 (to August)	77,271	53,999	59,991	191,261
	376,476	76,824	682,208	1,135,508

Apart from 1907, overseas business formed the majority of the value of work coming in to the firm from 1901, when the detailed records started, until the end of 1913.

The production of lenses and prisms was still largely done by outside contractors in 1907, with most of the simpler lenses bought from Chadburn Brothers in Sheffield and the more complex ones and prisms from Adam Hilger & Co. in London. The majority of optical work done at Glasgow was still connected with the building up of prism assemblies and the mounting of lenses into their cells for incorporation in the rangefinder bodies, rather than the grinding and polishing of optical glass into finished components. This was a difficulty that had vexed the firm since it began, and despite two previous attempts to achieve some degree of self-sufficiency through integrating the Adam Hilger business, Barr & Stroud remained dependent on remote suppliers who frequently failed both to maintain adequate quality and prompt delivery schedules.

The possibility of acquiring Adam Hilger & Co. Ltd reappeared in 1907 through the decision of the War Office to adopt the Marindin rangefinder. Being well-placed by the existing association with both the instrument and its designer to bid for whatever contract was offered, Hilger's managing director, Frank Twyman, moved to establish a collaboration with Barr & Stroud for production on an anticipated scale which his company could not undertake on its own.[67] Barr and Stroud had made most of the mechanical components for the experimental rangefinders that Hilger's had built, and had dealt personally with Marindin on several occasions. Having learned of the War Office's selection, Harold Jackson wrote to him in mid-July 1907 observing, perhaps a little sourly, that as it had been 'impossible' to work with the War Office, Barr & Stroud would in future concentrate on foreign armies. Soon afterwards, Twyman approached Jackson with the idea of tendering jointly for the expected contract, a proposal which Jackson accepted on 26 July but quickly set aside in favour of a plan to set up a completely new business. Less than a fortnight later a detailed draft proposal was sent to Twyman under which each firm would provide half the capital for a venture to 'erect and complete' rangefinders using mechanical parts from Barr & Stroud and optical ones from Hilger's.

The proposal, however, went further than merely arranging the division of provision of parts. Noting that it was 'to their mutual advantage to work in association', the seven year agreement provided that Barr & Stroud would not make the Marindin rangefinder on their own behalf and Hilger's would not make 'rangefinders or parts for rangefinders' for anyone except the new company and Barr & Stroud. The failure to fulfil any of the agreement's clauses by either party would incur the then very substantial penalty of £5,000. Despite Twyman's ready acceptance within a week, the contract was much to his disadvantage because it committed Hilger's to not making any rangefinder products at all on the firm's own account. Jackson's draft did not limit such prohibition to the Marindin, which precluded Hilger's, or Twyman individually, from benefiting from any opportunities of spin-off that the rangefinder might generate and absolutely prevented them from contracting to any other maker of rangefinders who might later appear. A covering letter Jackson sent with the draft hoped that the project would 'enable the intimate relationship ... between us to be continued', although one interpretation might be that Barr & Stroud was looking for enforced fidelity in perpetuity. During September Twyman began to have second thoughts and requested changes that would let him exploit any new skills or opportunities that arose, only to be rebuffed by Jackson who refused point-blank to give way. His letter on 2 October made it clear that Barr & Stroud would not continue to give Hilger's optical work while Twyman was free to act in any way that might 'seriously tell against [Barr & Stroud's] interests'. He closed by saying that he would nevertheless be 'best pleased' if a closer working arrangement could be achieved and invited Twyman's suggestions.

Between October and mid-November, proposals for a 'wider agreement' were prepared, but nothing had been signed either for it or the earlier idea when it became clear that there was not going to be any large order for the Marindin.[68] On 20 November Jackson congratulated Twyman on his War Office contract for just sixteen rangefinders, and although assuring him that Barr & Stroud would supply the parts, cast him adrift to handle dealing with the War Office alone, saying 'I hope you will be entirely successful'. Jackson then raised issues which suggest that despite the pleasantries Barr & Stroud was actually seeking to annex the Hilger business. Twyman was reminded that the failure to win a large order 'only postponed' the earlier proposal to form a joint production company which would be immediately implemented if a big contract eventually emerged, or if 'the suggested wider agreement' was not carried through. This left Twyman in a difficult situation, where failure to accept Jackson's latest proposals would require him to find half the capital for a new company and works which would subsequently have nothing to do except make sixteen rangefinders. If he demurred and sought to avoid that expense, then he, or Hilger's, would be liable to pay Barr & Stroud a very substantial penalty of £5,000 that would no doubt have left him and Hilger's seriously harmed financially. Jackson's assurance that the 'wider agreement' would not change either the name or the character of Adam Hilger & Co. Ltd was doubtless of little consolation to him.

Having brought Barr & Stroud to a position where it looked likely to gain practical control of Hilger's, Jackson's next action is surprising. With Twyman in a situation where he seemingly could do little but agree to whatever was being put to him, on 3 December Jackson wrote that Barr & Stroud now wished to withdraw from the matter and retain the status quo. His reason was that 'Mrs Hilger', the widow of the firm's founder and Twyman's fellow shareholder, might suffer if Barr & Stroud encountered poor trading which would, under the pro-posed arrangements, 'seriously affect Hilger's financial position' and even wipe out its profits completely. Barr & Stroud, said Jackson, had previously had years of trading at an 'absolute loss', by which he was referring to the early 1890s, and 'we would not like to contemplate such a contingency'. He ended with the com-ment that under the circumstances it would be better for the firms to remain independent 'unless you are willing to consider the sale of your business out-right, which I do not think likely'.

This curious outcome – a remarkable volte-face by any standards – has a likely explanation which relates to the firm's management structure and the characters of its owners. Although the business was a three-way partnership in which its two founders retained absolute control, Jackson was in effect its man-aging director who had since the mid-1890s been responsible for the way the firm conducted its business. He also believed that good management involved being willing to take firm, and not necessarily popular, steps to get things done.[69] His frequent letters to Twyman at Hilger's show that he was willing to make his feelings clear about poor workmanship and delivery times, on one occasion writing that 'We cannot express ... how strongly we have been annoyed by the way you have neglected us [over an order]'.[70] Nor was he the only one in the firm with a sometimes poor opinion of Hilger's. William Stroud also had strong feel-ings about their chief optical contractor's abilities, telling Barr early in 1907 that he doubted if Hilger's knew 'one hundredth of what Zeiss knows'.[71] Given Jack-son's philosophy of business management, his letters to Twyman suggest that he believed improvements could only be made by reconstructing Hilger's direction, which was solely under Twyman's control. The careful phrasing of the proposed agreement in 1907 indicates that Jackson intended to achieve a situation where Barr & Stroud would have management control, if not outright ownership of its principal optical contractor, a sentiment which might not have sat easily with either William Stroud or Archibald Barr.

Although Stroud's assessment of Hilger's optical abilities was less than favourable, he had never been able to suggest any viable alternative to using the firm. At the same time, he had previously resisted suggestions of any formal con-nection.[72] Barr, on the other hand, had long been acquainted with the now-dead Hilger brothers, both of whom he had respected and seen as friends, as well as having a largely amicable attitude towards Twyman.[73] Jackson's efforts to badger the latter into relinquishing control of Hilger's would have been unlikely to appeal to either partner – to Stroud because of his low opinion of the firm, and to Barr because of his previous personal involvement. Despite his energy and commitment to the business, Jackson lacked the power to force his principals into acquiescing in his proposals and the evidence suggests that he was obliged to extricate himself as well as he could from a situation that had escaped his con-trol, leaving the firm still with an unresolved problem about optical components.

Relations with Hilger's reverted to their earlier pattern, and not until 1912 did conditions force Barr & Stroud seriously to reconsider their arrangements for high-grade lens and prism work. By then, the large pentagonal prism end-reflectors which had become standard in the rangefinders were being flat-polished at Glasgow from pre-moulded blocks supplied by Chance Brothers in Birmingham, with nineteen glass-workers employed full-time.[74] Orders increased substantially that year and Jackson began to seek suppliers besides Hilger's, not just because of quality problems or limited capacity but also because the Admiralty had begun to suggest that Barr & Stroud should reduce its dependency on a single supplier.[75] He approached the Dallmeyer Optical Co., the Ross Optical Co. and Taylor, Taylor & Hobson Ltd for lenses, as well as ordering more from Chadburn Brothers.[76] In an effort to acquire a full complement of skilled workers and all the expertise needed to let Barr & Stroud attain self-sufficiency in spherical lens production, Jackson took the bold step of attempting to simultaneously recruit, or rather poach, the entire workforce of the Periscopic Prism Co. in London, a firm with which Barr & Stroud had been doing business since April 1911.[77] He told his contact there, Paul de Braux, that if Barr & Stroud could get an 'energetic and capable foreman' the firm would enlarge its optical shop and be 'quite willing' to employ as many skilled men as could be persuaded to leave the Periscopic Co. As de Braux was actually the firm's proprietor, he was understandably somewhat unwilling to give up his own business to be Jackson's energetic foreman and he declined the offer, leaving Jackson to carry on the still fruitless search.

The problem was made worse because there was no comparable work being done in Glasgow or anywhere else in Scotland and there were virtually no facilities for training in optical manufacturing outside London. Starting a spherical lens shop from scratch could not be done without a nucleus of skilled labour, and all efforts to tempt workers from England had so far been almost wholly unsuccessful. Then unexpectedly in November 1912, seven optical workers from Thomas Cooke & Sons Ltd approached Barr & Stroud and offered to move to Glasgow because work at the York factory had 'become slack' following Arthur Pollen's delays in selling his fire-control system to the Admiralty.[78] Welcome as the recruits doubtless were, the increasing level of orders throughout 1912 and 1913 meant that the lack of optical capacity continued to cause delays in output. Jackson added the London firm of W. Watson & Sons to his lens and prism suppliers and even began buying large pentagonal prisms from the German firm of J. D. Moeller in response to their unsolicited offer of supplies, presumably to meet overseas orders in view of the Admiralty's proscription of foreign glass.[79]

Barr & Stroud's efforts to integrate backwards into lens and prism production by 1914 were still not enough to make the firm independent of suppliers over whom it had little direct control. Had Jackson's moves to annex Hilger's been allowed to proceed in 1907, then the firm would have been able to develop its capacity either by wholesale removal of the factory and workers or, more likely, by the acquisition and transfer of technology and the savoir faire of craft technique that was still part and parcel of optical manufacture in the early twentieth century. Passing over the opportunity delayed that phase of development with the result that the firm failed to match optical capacity with that for mechanical engineering and assembly, a situation that, as the following chapter will show, posed serious (though not insuperable) difficulties when the First World War began.

The Admiralty's Other Optical Munitions Dealings

Although rangefinders accounted for the largest value in the Admiralty's optical purchases, they were by no means the only important ones. One other category was at least as significant to the long-term future of the industry, contributing to what were becoming identifiable as strategic weapons systems. The submarine periscope was a particularly important example of the Navy's adoption of new technology encouraging optical munitions manufacture.[80]

When the Admiralty acquired its first submarine in 1901, it was still an experimental vessel whose future remained unclear.[81] It was already recognized that to be a viable weapon the submarine needed some means to allow above-surface vision when it was submerged, but no practical device had yet been produced for any of those built.[82] Soon after the Royal Navy began investigating the possibilities of its first submarine, one of its leading proponents, Captain Reginald Bacon, proposed a design for one which was made up and installed in the first vessel. He and his prototype were then introduced by the submarine's builder, Vickers, to Sir Howard Grubb, the owner of the highly regarded astronomical telescope making firm in Dublin. Grubb had already shown some interest in optical munitions and had been granted British Patent 10373 in 1901 for an instrument 'facilitating the sighting of distant objects from submarine boats, barbettes and other protected positions'. This resulted in the firm becoming not just Britain's first maker of periscopes, but one of the earliest successful ones in the world.[83] Grubb redesigned Bacon's model and patented his improved version, going on to refine and develop it to a considerable level of sophistication.[84]

The subsequent fostering of the periscope industry was different to rangefinders and the Admiralty had little or no direct contact with Grubb, dealing instead with Vickers who delivered the vessels with the periscopes already installed. The numbers needed were far smaller than for rangefinders (only one for the earliest boats and two each in those built after 1905, with less than 100 submarines built before1914), but the later ones cost over £465 each and were comparable in price to a large Barr & Stroud rangefinder, so providing additional income for Grubb's telescope business. The monopoly of submarine building which Vickers maintained up to 1914, and the company's continued use of Grubb meant that his firm became, like Barr & Stroud, the de facto monopoly supplier. That state was threatened when serious service criticisms of the Grubb instruments drove the Admiralty to examine foreign instruments to find a superior type.[85] Because the Admiralty had then formulated a policy of non-reliance on foreign suppliers, once suitable types had been identified licences were taken out in 1911 from C. P. Goerz in Germany and Officine Galileo in Italy, by the Glasgow company Kelvin Bottomley & Baird.[86] This successfully established an alternative supply source, which presumably galvanized Grubb into making the necessary improvements to his designs. At the outbreak of war in 1914 periscopes made by both firms were being fitted to the boats being built by Vickers.[87]

The evolution of periscope supply had some things in common with the naval rangefinder's. In both cases the Admiralty was able to avail itself of a capable industry without any direct investment by the state, and to be assured of the regular delivery of instruments which kept pace with its shipbuilding programmes. But there were also significant differences. With the periscope, despite Grubb's early success, no export activity ever developed. That may have been because some measure of secrecy surrounded submarine building and Vickers were unwilling to allow Grubb to follow an independent marketing path. The scale of foreign submarine construction was far less than that of surface warships and the nations building them developed their own designs, France, Germany and Italy all having domestic periscope capacity well before 1914. Furthermore, Grubb's position vis-à-vis Vickers meant that his company was essentially a sub-contractor in the building of the boats, unlike Barr & Stroud who supplied rangefinders to the Admiralty for installation on all its new ships, irrespective of their builder.

By 1914, Barr & Stroud had become the world's largest specialist maker of optical munitions, and was far closer in character to the armaments industry than to the scientific instruments industry where it has usually been thought to belong.[88] The nature of armaments firms at the time has been characterized as having little resemblance to those who worked in the civilian marketplace and military contracting further described as an unwelcoming field far removed from the rest of commercial commerce.[89] Private enterprise rather than government departments had been responsible for the greater part of advances in weapons technology and the greatest talent of the arms industry lay in its ability to combine such disparate skills as heavy engineering with the finest high precision engineering work. That exactly describes the kind of work done by Barr & Stroud and by Grubb, who fitted prisms as small as an inch and lenses less than half an inch in diameter made to tolerances of less than one ten thousandth of an inch into massive metal structures up to thirty feet in length and so heavy that they required lifting gear to move them around the assembly shop. Even if large rangefinders and periscopes were not wholly typical of optical munitions manufacture, other instruments like the dial sight nevertheless conformed equally well to the characterization of armaments, in that they had no civil market and their demand was inseparably linked to the state's needs at any time, so that attempts to sell them were constrained by factors quite different to those in civil markets. By the beginning of August 1914, Britain had evolved an optical munitions industry that had the capacity to supply not only all the nation's own peacetime demands, but many of those of foreign powers besides. It was a numerically small but nevertheless distinct industry specializing in the manufacture of optical goods for the armed forces that the majority of optical companies were unable or unwilling to tackle. Most of the constituent firms produced military optics in addition to, but separately from, other unrelated commercial products, deriving only a proportion of their incomes from government contracting, but the largest one relied entirely on the international demand for armaments to provide its business. None of these firms had been given any state assistance in developing what they made, but neither had they been discriminated against as has been previously suggested. The peacetime armed forces most certainly did not rely on imported instruments; they could, and did, draw their whole requirements from an independent and generally capable domestic industry that was geared to contemporary levels of demand.

The British optical munitions industry had reached what amounted to a state of equilibrium in meeting the needs of the Admiralty and the War Office by mid-1914. That was to be completely upset by the challenges and problems that followed the declaration of war in August 1914 and the story of optical munitions now moves on to deal with the industry's trials and tribulations over the ensuing five years.

4 THE IMPACT OF WAR, AUGUST 1914 TO MID-1915

The optical munitions makers, like almost all of British industry, were unprepared for the demands of a major war and encountered a variety of problems in responding to the growing demands made upon them between August 1914 and the early summer of 1915. Some of these difficulties were outside the industry's own control, but others resulted from the overall structure of optical instrument making within which most of the established munitions contractors lived. The principal difficulty was the unanticipated, exponentially increasing demand from the British Army that was expanding on an unprecedented scale and which had not yet been fully equipped with optical apparatus before the war. In the first ten months of the conflict, the War Office failed not only to quantify its own optical requirements accurately, but also neglected to concentrate the orders it did place on the makers who were best suited to deal with them. There was an inability to recognize, let alone come to terms with, the strengths and limitations of both the general and specialized optical sectors of the industry. Its strengths were either ignored or disregarded during 1914 and early 1915, and the ensuing shortcomings in deliveries have been read as signifying a chronic systemic weakness in the industry, particularly in that it had failed to keep up with both its French and German counterparts. In fact, the difficulties in optical munitions output were more deeply rooted in the problems of industrial mobilization than in any backwardness of the industry. The procurement process was, initially at least, by far the greater obstacle to output.

The War Office had become a steady, though far from prolific, buyer of optical munitions before 1914. Its purchasing had been influenced by the combination of indecision in deciding on suitable patterns and the phasing in of new weapons and budgetary constraints, rather than a failure to appreciate their importance in its armoury. Contracts sufficient to equip the regular forces with prismatic observation binoculars and the complex artillery dial sight had actually been placed but were so scheduled that deliveries were still not fully complete when war was declared. Orders for infantry and artillery rangefinders had only recently been finalized and hardly any of what was needed had been delivered by August 1914. Most of the problems met by the optical munitions makers in the first year of war were in trying to satisfy the War Office's rapidly developing needs. The Army's requirements were geared directly to its numerical strength which grew enormously and at great speed during the late summer and autumn of 1914, in contrast to the Navy where demand was linked to ship construction and increased relatively slowly. Emphasizing these differences, the Admiralty had been a regular and consistently substantial purchaser of optical instruments over the preceding decade and had established an effective liaison and an efficient supply chain with makers who had become familiar with its needs. As a much smaller and less regular client, the War Office had still to reach the stage where it valued, or was valued by, the optical munitions community to the same extent.

The Capability and Capacity for Optical Munitions Manufacture

Even before the problems which emerged in late 1914, there had been concerns outside the armed forces about their ability to get the optical apparatus they would need in the event of war. These worries originated not from the munitions sector itself but within the broader community of optical and scientific instruments makers and users. They focused on the British optical instruments industry as a whole, which was seen by some as inadequate to meet possible requirements should Britain be drawn into a European war. In June 1914, the British Science Guild considered that under such circumstances British firms 'could not, unaided, produce sufficient quantities' of the optical devices that would then be wanted by the armed forces.[1] These sentiments tended to echo a more general concern already voiced by some members of the general optical industry, focusing on the notion that its abilities could be improved by the provision of formal education and training facilities.

According to the account subsequently left by the Ministry of Munitions which was created in the spring of 1915, the capacity for supplying optical instruments to the British Army in late 1914 soon became a matter of grave concern. The Ministry's official history later described British optical manufacturing as then being both 'seriously undercapitalized' and 'very conservative', with 'such machinery as existed [being] antiquated'.[2] There was also 'a singular lack of comprehension' of the benefits of machine tools, implying a serious or even chronic lack of capacity to manufacture quickly and in quantity. That gloomy depiction has been echoed by later writers who have considered that the industry was 'a fragmentary collection of craft based family firms' which suffered from a 'scarcity of capital for investment and research'.[3] Another view was that once the war began 'it was vividly brought home to the British government ... that they were heavily dependent upon Imperial Germany both for finished precision instruments and for many component parts'.[4]

The reality was somewhat different. In mid-1914 there was already a viable British optical munitions manufacturing base which was more than adequate for all the requirements of the peacetime armed forces. Its members were well able to manufacture high precision complex military optics, progressively minded and far from antiquated. Besides the already experienced munitions contractors there was also a far from insubstantial industry producing a wide range of civil optical equipment which might be called upon to produce some, though certainly not all, of what the War Office now urgently needed. The civil instruments sector included at least twelve manufacturing companies of varying sizes who were regularly producing a range of products for science, surveying and navigation.[5] With just one exception, what neither sector was equipped to do was to increase output massively and at short notice.

The War Office's problems in providing the Army in France with sufficient optical stores in 1914 and early 1915 cannot be laid solely at the industry's door. The difficulties resulted from the combination of a totally unprecedented scale of demand, shortcomings in the War Office's procurement mechanisms and limitations in the optical munitions sector which were principally – though by no means exclusively – a consequence of the Army's pre-war philosophy about the constitution and tactical employment of its optical inventory.

The first stage in the industry's wartime expansion was largely concerned with increasing its output to meet the growing demands from the armed forces. The pressure exerted on the makers from August 1914 to the summer of 1915 was principally to produce more of what was already being made. Problems in organizing production occurred because the Admiralty and War Office competed for manufacturing capacity without any coordination, each exerting pressures on makers to satisfy its own requirements whilst apparently ignorant of the conflicts of interest sometimes created. By far the greater pressure came from the War Office, particularly when the need for new types of optical munitions began to emerge early in 1915 in response to specific requirements highlighted by experiences in combat. One particularly telling aspect of the War Office's limited ability in organizing development and production was the way in which contracts to develop new types of optical munitions were passed to firms who lacked the expertise to design instruments suitable for large-scale manufacture, compounding delivery problems still further. There was, as has rightly been pointed out, little in the way of coordination of research, design or output during this stage.[6] The optical industry – unlike, for example, the small arms industry – was left entirely to its own devices in organizing its industrial mobilization.[7]

The Procurement Problem: War Office versus Industry

Although the Army was no stranger to optical munitions, its pre-war use of them had hardly been lavish, not least because its own scales of issue did not call for their widespread deployment. In the financial year to 31 March 1914 less than £49,000 had been spent.[8] Irrespective of the small value of recent business, there was nevertheless a core group of makers thoroughly experienced in producing specialist instruments to War Office requirements. Those included strict conformity to specifications and quality at a level not required in the civil markets where all the contractors except Barr & Stroud also competed. In the preceding three years, ten British companies had supplied binoculars, dial sights, rangefinders, sighting telescopes and signalling telescopes to the War Office.[9] There was also an eleventh manufacturer in the form of the German Carl Zeiss company's London branch factory set up in 1909 to produce prismatic binoculars principally for sale to the War Office.[10]

The despatch of the Expeditionary Force to France started a process which led to the creation of an unfavourable image of the optical munitions industry in the light of its alleged lack of success in meeting demands. It is undeniable that during the first years of the war there were indeed serious shortages of optical munitions, but the reasons for those deficiencies were as much to do with the War Office's Contracts Directorate as with the optical industry.

The War Office faced two problems in equipping front-line units. The provision of up-to-date optical apparatus for the peacetime regular Army was still far from complete because some of the most important orders had only recently been placed and the earlier contracts described in the preceding chapter were also still being filled. The initial shortages resulted not from any failure on the makers' part but from the timing of War Office orders, which had been placed without any anticipation of the declaration of war. The second difficulty compounded the first because the immediate calling-up of reserves drew into the British Army large numbers of men for whom no provision of up-to-date optical munitions had ever been planned.

The peacetime strength of the British Army was approximately 234,000, of whom 192,000 were front-line troops.[11] From these, an Expeditionary Force of 150,000 was meant to be despatched to the continent. The domestic Territorial Force of 256,000 men was to be mobilized to provide the defence of the British Isles.[12] The only optical munitions available for the latter were obsolescent ones which had earlier been released from the front-line units as new patterns came into service, together with those left over from acquisitions made during and immediately after the Boer War. No other provision had been made because no pressing need had been recognized and because funds were more urgently needed to supply front-line units. Then, there were additional reserves of 200,000 men of whom 56,000 formed a 'Special Reserve' intended to provide the Expeditionary Force with replacements and reinforcements. These soldiers needed the same equipment as the Regulars, but once again no provision had been made for optical munitions. The Army thus had the immediate prospect of mobilizing for front-line service 150,000 men who were inadequately supplied with optical stores, plus another 56,000 who were totally without up-to-date instruments. Besides those, another 400,000 men in the Territorial and general reserves almost entirely lacked optical equipment of any sort, constituting a problem that would have to be addressed if they were later to be committed to action.

The supply problem almost immediately worsened as it became apparent that a much larger field force than 150,000 was going to be required.[13] On 6 August Parliament approved the provision of an extra 500,000 men, of whom 100,000 enlisted before the end of the month. A further half million was voted for only five weeks later and before the end of November approval was given for another million, increasing the Army's pre-war strength to around 2.5 million. By the beginning of November, enlistments of one million since the declaration of war were starting to impose a massive burden on the supply of all types of munitions. All these factors constituted a recipe for chronic problems in optical munitions supply.

The scale of the Army's need for optical devices had, like everything else, rapidly multiplied beyond anything ever envisaged.[14] The head of Woolwich Arsenal's Optical Inspection Department quantified the Army's requirements as they stood in November 1914.[15] His department was responsible for inspecting all optical stores and he was fully conversant with their scales of issue. These meant that excluding gun sights and signal telescopes, 7 per cent of the Army would require binoculars and every 100 men would want a rangefinder. In mid-November, the Army's establishment had reached a million, so that 70,000 binoculars and 10,000 rangefinders were already required in addition to what the pre-war Regular Army and reserves still needed. The provision for another million men meant the figures would double by the following July when those new soldiers had all been inducted, making a total shortfall of 140,000 binoculars and 20,000 rangefinders, excluding telescopes and gun sights.[16]

The responsibility for procuring all munitions supplies lay jointly with the Master General of the Ordnance (MGO) and the War Office Contracts Department. The historian of the Ministry of Munitions provided an explanation of their respective functions which can hardly be bettered:

> The size of the Army being determined by Parliament, and the scale of equipment being approved, the formulation of definite requirements was a straight-forward matter. It was the duty of the Master General of the Ordnance and his officers to prescribe what equipments should be supplied and the duty of the Contracts Department was limited to procuring from the armaments firms such portions as might be definitely requisitioned.[17]

The MGO should therefore have been able to calculate what 'definite requirements' in optical stores were needed according to the growing size of the Army,
and then issue instructions to the Contracts Department to procure the quantities required. Following peacetime practice, the latter should then either have
offered contracts or have sought tenders from manufacturers before awarding
them.[18] Because the instruments needed in late 1914 were all of existing patterns,
and because established and recent sources of supply already existed, it might be
expected that sufficiently large orders would have been placed at then-current
prices to meet the immediately emerging demands. In fact this was not done,
so that by mid-1915 there was both a substantial and growing deficit in orders
to meet the growing demand and a shortfall in deliveries of optical munitions
already required.[19] The pattern of ordering was as much, perhaps even more,
responsible for the failure of the industry to provide adequate supplies as were
the shortcomings in its structure and background suggested by the Ministry of
Munitions. Unlike rifles, small-arms ammunition and artillery shells, where the
ever increasing and massive orders eventually led to the re-organization of the
supplying industries, large contracts for optical munitions were, with one important exception, never placed.[20] This was the exact reverse of what happened with
most other types of munitions.[21]

The problems of industrial mobilization certainly applied as much to the
optical sector as anywhere else in the armaments industry. The requirements of
warfare in 1914 did indeed create 'unique problems of supply', with complex artefacts needing to be made to consistently high standards of precision replication
in quantities not normally required in peacetime.[22] Other than in wartime such
items are usually made by only a small number of specialists but the pressures of
war create the need for their products to be turned out in the greatest numbers
possible and with the utmost haste.[23] This was the situation which the optical
munitions makers faced during late 1914 and early 1915, in circumstances certainly ripe for the creation of difficulties in coping with new demands. However,
the underlying inadequacies of production capacity were obscured by the signal
failure of the MGO and Contracts Departments to place the necessarily large
orders fully to meet current demands. Unlike artillery and ammunition supply,
where makers were persuaded to accept extremely large orders often against their
better judgement, the optical munitions suppliers were never asked to produce
anything like the quantities of materiel actually needed. Because of that, the optical industry's actual capacity to supply war demands was never properly examined
until mid-1915 when it fell under the aegis of the Ministry of Munitions.

At first glance, the failure to place adequate orders may seem inexplicable.
After all, the scales of issue were known, as were the rate and extent of the Army's
planned expansion. Nor were the items novel, and a list of 'approved contractors'
already existed, amongst whom orders could have been distributed. Further-

more, funding was available to cover whatever munitions stores the MGO put out to contract. The reasons why large-scale ordering across the industry did not take place were by no means solely to do with a failure properly to recognize the scale of requirements but were also substantially to be found in the operating system imposed on the Contracts Department by the regulations governing the organization of the War Office as a whole.[24]

The entire War Office procurement system was hedged around with safe-guards to protect against exploitation and default by contractors, and to ensure that the Army obtained goods of consistently good quality which conformed strictly to specifications and for which the state paid no more than necessary. In practice, this meant that priority was to be given to the lowest price tendered from an already approved supplier. To acquire that status, a manufacturer had to apply to go on the 'approved list' and be vetted to ensure the capability of carrying out the work involved. The MGO's office decided on the total quantities to be procured and then issued instructions to the Contracts Department to organize deliveries. The latter then sent 'invitations to tender' only to those who were on the approved list. The system had worked well enough in peacetime but had serious defects in war which were later explained by the civilian head of the Contracts Department, U. F. Wintour.[25]

His explanations highlight the mindset ingrained in those responsible for munitions procurement and go far to explain the inadequate levels of ordering for almost all optical munitions between late 1914 and mid-1915 when the Ministry of Munitions took over. Understanding the constraints under which the Contracts Department operated helps to explain the allegedly poor performance of the optical industry in this period. Wintour asserted that being forced by high levels of demand to obtain tenders simultaneously from all on the approved list had 'several vicious consequences' for the War Office. These included revealing to the supplying industry the actual scale of the Army's needs and letting the industry see for itself 'the relation of [Army] demand to the probable supply'. The urgency in 1914 and 1915 meant that 'all or most offers' had to be accepted, so that there was absolutely no chance 'to keep prices down to a reasonable level' by refusing tenders considered too expensive. Because one of the department's chief obligations was to secure the lowest practicable prices, attempts were made to force them down. This was done at first by issuing requests for tender in quantities lower than actually needed. That proved counterproductive because bidders quoted higher unit prices reflecting the lost economies of scale on very large quantities. Another ploy involved asking for prices based on whatever quantity could be offered by a specified date. That, according to Wintour, created an impression of a potential demand larger than it really was, and encouraged the makers to keep prices at what the Contracts Department considered a high level.

Rapid procurement was subordinated to the perceived need to prioritize price management and the department seemingly worked from the premise that contractors would overcharge in the absence of any control mechanism. Whether the optical suppliers actually sought to maximize their prices is uncertain. The records of Barr & Stroud, which are the only detailed ones available for this period, suggest that prices only increased in line with actual costs of raw materials.[26] Surviving War Office contract records from 1914–15 give no indication of prices so it is impossible to know whether such concerns of excessive charges were really justified.[27]

Another counterproductive effect of these attempts to manipulate market forces was the effect on the prices and apparent availability of raw materials. Wintour said that when all the firms on an approved list were simultaneously asked to tender for large quantities they all 'went into the market at the same time for the [necessary] raw materials', by which he meant that they took options on what they might need. In consequence, the apparent demand 'multiplied several times over', causing 'complete chaos in the market and [forcing up prices] to quite fictitious and unwarranted levels'. Within the context of optical munitions, this certainly took place with optical glass, the price of which rose considerably and which was periodically in very short supply during 1914 and 1915, delaying and disrupting output.[28]

For the Contracts Department there was neither an open cheque book nor any relaxation in the procedural system despite earlier very clear signals from government to the MGO's department that changes were necessary to speed the machinery of procurement. The tendering system had already been effectively made redundant in October 1914 when Lloyd George, then Chancellor of the Exchequer, had given the Ordnance Department 'virtual *carte blanche* approval' for the purchase of munitions stores,[29] but little difference became apparent at the Contracts Department. The existing system of competitive tendering and its now largely redundant safeguards for the public purse continued, even after the Ministry of Munitions was created in 1915. As late as February 1916, the Contracts Department complained that conditions of supply had become such that there was 'no real competition' between prospective suppliers, so that even the allocation of orders according to price could not ensure 'that the Department gets the best value for its money'.

The tenacity with which the Contracts Department held on to its peacetime procedures suggests that the tendency of the military society to show inertia and resist changes challenging a well-established and seemingly satisfactory routine was by no means confined to a technological context.[30] Wintour, although a civilian, was summarizing the Contracts Department's practices prior to the changes caused by the creation of the Ministry of Munitions which effectively took over the placing of contracts once requirements had been produced by the

War Office. Despite the then-manifest shortcomings in earlier procedures, and the political will to disregard the emphasis on competitive tendering, he was emphasizing that his department considered its problem to be essentially that the market was failing to conform to previous expectations, and that what had been perceived as necessary by the Contracts Department was some means to re-establish the earlier familiar stability in the procurement process. Whether or not institutional inertia was all that encouraged the Contracts Department to maintain the status quo, there is no doubt that the tendering system had an unfortunate effect on optical munitions procurement. That the Contracts Department's vision of fiscal prudence was not unique was evidenced in the month before Lloyd George sanctioned the relaxation of procurement procedures, when the MGO had deliberately concealed from the armaments industry the provision of £20 million by the Treasury for plant extensions, 'fearing a great onrush of supplicants for the funds' and because he was generally reluctant to sanction large-scale expenditure.[31]

The net result of all this for the optical munitions suppliers was that, apart from Barr & Stroud, they were never asked in 1914 even to attempt to provide sufficient capacity to cope with the massive demands for the growing Army. The case of Barr & Stroud is particularly interesting because it shows simultaneously some of the strengths and weaknesses present in the early wartime industry, and suggests that had the War Office approach to ordering been modified to reflect the needs of the Army rather than the desire to preserve an existing administrative structure, then the Ministry of Munitions might well have found an altogether different picture to the one it claimed to discover in mid-1915.

Barr & Stroud's Experience in the First Year of War

By August 1914 Barr & Stroud was the world's only company devoted exclusively to the manufacture of optical munitions and had become Britain's largest producer of high precision optical apparatus. At home, the Admiralty depended wholly on the company for its rangefinders, and abroad the firm had achieved a monopoly in supplying rangefinders to the navies of France, Italy and Japan, as well as a hegemonical position in virtually every other modern navy except those of Germany, Russia and the United States of America. Its exports of rangefinders to the world's armies were also very substantial, with France being the largest single client between 1912 and 1914. This success had been achieved without the benefits of either encouragement or substantial orders from the War Office before 1914 and, as described in the preceding chapter, there was no harmonious or even well-established relationship between the company and the Army.

Barr & Stroud was very much an export-orientated business and had been dependent on foreign orders for much of its profitability since 1901. In its financial year ending 31 December 1913, 64 per cent of the turnover of £268,000 came from overseas business, compared with 34 per cent from the British Admiralty and only 2 per cent from the War Office.[32] Profit margins on overseas business were higher than on similar items sold to the British forces; infantry rangefinders for France were sold at £66.50 when the same item was priced at £55 to the War Office.[33] The factory's output was divided between large naval instruments, many of which were for the Royal Navy, and the smaller portable models for land service which had been sold almost entirely abroad. For these, the principal customers had been European armies, with France being by far the biggest client with over £100,000 of orders in 1913 alone. It was only in the spring of 1914 that Barr & Stroud received its first large War Office orders, an event that proved to be something of a mixed blessing for the firm.[34]

Those contracts were not awarded through the usual competitive tendering process because the firm was considered a monopoly supplier and thus exempt from the general rules governing the placing of War Office orders. No other British maker could supply anything similar in quantity, and the only substantial foreign manufacturers of rangefinders, C. P. Goerz and Carl Zeiss, were both German. The uncertain political climate in Europe even before mid-1914 meant there was no possible source other than Barr & Stroud and the Contracts Department's usual duty to obtain best value was accordingly redundant. The first War Office orders were certainly large, but by no means unusually so for Barr & Stroud. Between January and July 1914 four separate contracts were placed for £54,000 worth of rangefinders, compared with over £185,000 of foreign business and £73,000 from the Admiralty. The War Office orders still did not represent a complete outfit of instruments even for the Regular Army's peacetime frontline infantry strength of 150,000 men; at least 1,500 were needed but only 680 were ordered. Similarly, only 150 were ordered for the whole of the field Artillery. The declaration of war prompted an increase in ordering, with new contracts placed before the end of December for £119,000-worth of instruments, followed by orders for over £300,000 by the end of July 1915. These later contracts were indeed very large. The firm's greatest annual peacetime orders had totalled £268,000 in 1913, but by the end of 1914 new business came to over £491,000, an unprecedented increase which, according to the Ministry of Munitions, was handled so efficiently that rangefinders were the only optical stores whose deliveries to the Army were 'assured' right from the outbreak of war.[35] However, that assurance of rangefinder supplies actually resulted from a combination of fortuitous circumstances and the firm's ability to manage a diverse set of problems, some of which were actually created by the War Office itself.

In one way, the War Office was the fortunate, if not entirely deserving, beneficiary of Barr & Stroud's earlier marketing efforts in Europe. Whilst the British Army was still deliberating as to what rangefinder to adopt, the firm had been busy selling on a very large scale on the continent. Growing political tensions in central Europe and the Balkans had encouraged continental powers to complete the general updating of military equipment which was already taking place and Barr & Stroud had procured particularly large orders from France, prompting the company to increase its plant and equipment to handle not only that extra business but also large anticipated orders from Austria–Hungary and Russia. By mid-1914 the first contracts from the former had been signed, and negotiations with the latter were nearing completion.[36] The Austria–Hungary order had been won after an international competition in which every rangefinder-making company in Europe had competed (see Figure 4.1).

Figure 4.1: Competing infantry rangefinders in the Austria–Hungary army trials. Illustration courtesy of University of Glasgow Archive Service, GB0248/UGD295.

Manufacture had actually begun by August, but the declaration of war saw the client immediately become an enemy and also put the Russian negotiations into abeyance. These contretemps, and the co-incidental near-completion of the large French contracts, meant that sufficient capacity was available to handle the new War Office contracts placed after the outbreak of war. Without the existing tooling and machinery from those pre-war orders, and the concomitant availability of skilled labour, the immediate and adequate deliveries of rangefinders for the British Army would have been neither assured nor easy.

Easy was, however, very much a relative term in the context of manufacturing in late 1914 and early 1915. Factory space, machinery and labour, although essential, by no means defined the limits of manufacturing problems. Despite making optical instruments, Barr & Stroud regarded itself primarily as a mechanical engineering company. The bulk of its workshop space, machinery and workforce were allocated to fabricating the mechanical structures of rangefinders (see Figure 4.2).

Figure 4.2: Barr & Stroud naval rangefinder assembly shop immediately before the outbreak of war, showing 9-ft and 15-ft models. Illustration courtesy of University of Glasgow Archive Service, GB0248/UGD295.

Optical components were at the instruments' heart, but it also depended equally on skilled and precise mechanical engineering for its efficient working. Barr & Stroud had become entirely self-sufficient in the mechanical side in the previous decade, but still depended heavily on subcontractors for many of its most important optical components. The way the business sought to manage the sudden increase in requirements for them provides a useful illustration of the ability of the optical industry as a whole to cope with wartime demands in 1914.

Although Barr & Stroud rangefinders all shared a basically similar optical design, the different models required their own outfits of lenses and prisms, depending on the base-length and magnification. Some were made in-house, but many were still being done by outside contractors who frequently failed to reach the specified standards needed for the instruments to work properly. Barr & Stroud's problems with increasing output in response to demands for rapid delivery were largely centred on maintaining the quality of externally sourced components from suppliers over whom there was no direct control. The scale of demand for lenses and prisms was now governed by the requirements of the two British armed services, which in turn influenced the extent to which Barr & Stroud had to depend on contractors who were all over 200 miles distant from Glasgow and very much preoccupied with their own increasing levels of business.

Once war was seen as inevitable, the directors began to put themselves on what they called a 'war footing'. On 31 July they assured the Admiralty that priority would be given to Royal Navy contracts over foreign ones and, apparently, over War Office ones as well.[37] On 3 August, general manager Harold Jackson wrote to all the company's optical subcontractors urging immediate attention to 'all our orders in hand'.[38] When war was declared the following day he sent the Admiralty a detailed list of all rangefinders on order, for both allies and enemies alike, and specifically requested an assurance that there was no objection to continuing supplies to France. The company had accepted that the Admiralty had the ability to direct exactly what it could and could not do. The Admiralty's wishes were important because they stood to affect the firm's requirements for optical components from outside suppliers. Having been assured on 6 August that there were no objections to deliveries for the French army and navy, Jackson began attempting to arrange for adequate supplies of lenses and prisms.[39]

The correspondence in the following months illustrates the problems that the optical munitions makers had to deal with, and shows that irrespective of capacity or skill there was indeed a lack of organization in the optical industry. The firm's efforts were ultimately successful, not through obtaining larger deliveries from existing suppliers, but through the expansion of internal capacity and the establishment of a new relationship with one particular maker.

Barr & Stroud's oldest optical supplier was Adam Hilger Ltd, and it was their Managing Director, Frank Twyman, that Jackson first approached. The two companies had been associated since 1889, and on at least two previous occasions a merger had been discussed.[40] Nothing had resulted, and by 1914 Hilger's – although still the principal British maker of complex prisms – had lost the near monopoly of Barr & Stroud optical components it once had. Hilger's had felt the diminution of Barr & Stroud orders very keenly after 1913 when the effect was said to have been extremely serious.[41] This reduction may partly have been because Barr & Stroud was becoming more self-sufficient, but Twyman later considered the real cause was Admiralty pressure on Barr & Stroud to insure against the dangers implicit in the possible failure of one key subcontractor, which had led the company not only to expand its own optical output but to engage other suppliers as well.[42] Whilst the Admiralty certainly was aware of dangers in interruptions to its supply of optical munitions, it is equally possible that the stimulus for self-sufficiency may have been encouraged by Hilger's continued inability to maintain sufficiently high and consistent standards. Although orders were still being placed for lenses and prisms in 1914, complaints of poor quality were frequent. Nevertheless, it was on Hilger's that Barr & Stroud at first intended to rely for the expansion of deliveries.[43]

Scarcely a week after the war began, Jackson asked Twyman for delivery dates for everything on order. Jackson emphasized that the size and the new urgency of existing orders meant Twyman must hire as many optical workers as possible, and visit Glasgow to discuss the situation with Barr & Stroud's directors. That letter, dated 10 August, must have been written knowing that an urgent War Office order for 558 infantry rangefinders was already on its way to Glasgow.[44] A meeting was held on 21 August, when the firm presented Twyman with a proposition he may well have felt unable to refuse.

Barr & Stroud proposed to give Hilger's all the optical orders that they 'could not undertake themselves'.[45] The directors knew these were greater than Hilger's own capacity and understood that Twyman would have to subcontract many of them. As Barr & Stroud was already outsourcing a substantial amount of optical work to firms besides Hilger's, the offer suggests that either the directors wished to simplify their own administration or, probably more likely, were aware that their existing subcontracting network would very soon be insufficient. Twyman would be welcome to subcontract, provided he took responsibility for the prompt delivery and quality of the work.

The directors also told him that because Hilger's was 'at the centre of the optical trade' he would be 'probably better able to subcontract [and] control the quality of the work than we are'. To assist Twyman in what was likely to be a Herculean task, they would help to obtain the necessary optical glass from the sole British makers, Chance Brothers of Birmingham, as well as providing advice on manufacturing methods. This was, at first sight, a substantial vote of confidence in Hilger's which would have left Barr & Stroud free to concentrate on its own manufacturing without the burden of coordinating the work of numerous subcontractors. It was, however, still short of a total endorsement and by no means gave Twyman any guarantee of subcontracted work because, irrespective of the proposed agreement, Barr & Stroud reserved the right to extend its own optical department as it thought necessary. Jackson's summary of the meeting noted that 'for the present neither you nor we can undertake [unaided] the work in hand', perhaps implying that he saw no reason why his firm should not move closer to that condition.

Barr & Stroud may have meant to flatter Twyman by saying that Hilger's was 'at the centre of the optical trade', but the statement was literally correct. In 1914 London remained the principal locus of the optical instruments industry. Apart from Barr & Stroud, the only substantial companies not based in London were

Thomas Cooke & Sons Ltd in York and Sir Howard Grubb & Co. in Dublin, both of whom were directly involved in optical munitions. There were other optical makers in the provinces, but none of them were large or substantially engaged in manufacturing for the armed forces. The London businesses ranged from relatively large and successful ones such as the Ross Optical Co. Ltd which employed over 350 workers,[46] down to small firms which were mainly engaged in ophthalmic optics and sometimes employed only one or two workers.[47] There was a concentration of makers in the Clerkenwell district where Hilger's was located, so that Twyman was indeed conveniently placed to liaise with many optical manufacturers, although his competence to do so was not so readily apparent.

Twyman's interests lay principally with the development of new types of optical instruments for science, particularly the spectrometer.[48] His company's other main area of expertise was the making of complex prisms, in which it had long held something approaching a monopoly in Britain. Both were essentially small-scale operations reflecting contemporary levels of demand, so that Hilger's remained a small business of less than thirty workers and Twyman had no experience either of volume production or the organization of subcontracting on the scale that was now necessary. Barr & Stroud's directors would have known this, and it is unclear why they so readily planned to devolve the task to him. The most probable explanation is that Twyman, through his prism work, was indeed well acquainted with most, if not all, of the London firms who regularly did the high-grade spherical lens work used in the scientific instruments that employed Hilger's prisms. These firms, few of which Barr & Stroud had ever dealt with, were now likely to become useful suppliers, if they could be persuaded to take on the work. Personal persuasion and cajolery might have been recognized as more effective recruiters than Jackson's sometimes peremptory letters and telegrams from Glasgow.

Whatever Barr & Stroud's thinking, by the end of October it was apparent that Twyman's efforts were not going well. Although none of Hilger's surviving records relate to this, Barr & Stroud's letter books document the problems and the frustrations they caused. Despite offering to make Hilger's responsible for all orders beyond Barr & Stroud's own capacity, the firm not only excluded the precision engineers and lens makers Taylor, Taylor & Hobson Ltd of Leicester from Hilger's remit, but also continued to deal directly with three London firms who were already supplying parts. R. & J. Beck, the Periscopic Prism Co. and W. Watson & Sons all had contracts which continued to be administered from Glasgow subsequent to the arrangement set up with Twyman.[49] Nor did Barr & Stroud leave Twyman alone to get on with his task; Jackson repeatedly intervened with instructions as to how the subcontracting should be organized.

Barr & Stroud's instructions to Twyman to exclude Taylor, Taylor & Hobson resulted from Archibald Barr's acquaintance with one of its principals, Wilfred Taylor, and his high opinion of the firm's workmanship and methods that dated

from 1903.[50] The Leicester company moved into lens making as part of a programme of diversification from the manufacture of small precision machine tools
and doing fine mechanical work. It had begun contracting for the War Office
from 1910 and by 1914 had expanded its optical side into the series production
of high-quality photographic lenses to designs licensed from Thomas Cooke &
Sons Ltd of York.[51] On 9 August, Taylor offered Archibald Barr his firm's assistance in the production of rangefinder object lenses which was quickly accepted
after a meeting of the two firms' directors.[52] Whether this was a deliberate hedge
against Hilger's failure is uncertain, but in view of how the other subcontracting
exercise turned out, it was certainly a propitious move.

Taylor told Barr & Stroud on 25 August that he had received third-party
enquiries for quotations for lenses to be supplied directly to Hilger's. Then, on
2 September , he was asked by Cooke's of York to quote for objective lenses to
be delivered to Adam Hilger & Co. Their specification identified them as being
for Barr & Stroud rangefinders and Taylor consequently declined the business.[53]
Cooke's were one of the country's best-known lens and telescope makers, and
were not currently engaged on any large-scale British government contracting.
The referral of the order to a third party could hardly have been because of a lack
of ability, but was more likely because they were still attempting to sell rangefinders of their own design intended for use with the Pollen fire-control system
to the Admiralty and the Russian governments, as well as promoting smaller versions for land-service use.[54] Whatever Cooke's reasons or motives, Taylor's letter
warned Jackson that if Twyman's subcontractors were passing on work, it could
pose threats to maintaining quality control, and he promptly gave Twyman specific instructions to tell them that they were not to delegate orders on to a third
party.

This division and subdivision of manufacture highlights a problem facing
firms seeking to increase optical output in late 1914 and early 1915. The capacity
of most optical manufacturers to expand production rapidly and substantially
was constrained by the structure of the pre-war industry which had operated on
a scale geared to domestic and export demands that generally lacked urgency. The
optical munitions component of the larger optical instruments industry had no
significant peacetime problems in meeting the state's demand, whether through
in-house manufacture or outsourcing. Even where the scale of demand had been
consistently large, as with Barr & Stroud's business, the required delivery rates
had usually been leisurely enough to allow supplies to keep up with schedules.

The expansion of orders and the need for increased speed of output strained
the companies to whom Barr & Stroud and Hilger's immediately turned. Beck,
Cooke and Watson, for example, were already producing optical munitions
components as well as complete instruments but were not wholly devoted to
those fields; they also had a substantial involvement in the general commercial

market for optical goods which they were initially loathe to relinquish. Increasing output for Barr & Stroud meant either reducing the output of something else or investing in new machinery and labour. Early in the war there was a general reluctance to do this when its duration – and hence commercial value – was expected to be short rather than long. That sentiment was reinforced by the lack of any avalanche of orders from the War Office to counteract it. Faced with such uncertainties and the lack of any centralized direction, it is hardly surprising that firms attempted to cope with what they saw as a short-term phenomenon by equally short-term measures such as further subcontracting.

For Barr & Stroud, the merit of that reasoning was subordinate to the firm's immediate problem of obtaining enough components of a sufficiently high standard. Throughout September Harold Jackson complained to Twyman about slow deliveries and erratic quality.[55] By 14 September, he was so worried about Hilger's prism output that he urged Twyman to approach the Periscopic Prism Co. (with whom Barr & Stroud was already dealing) for extra supplies of the rangefinder pentagonal prisms which had previously been one of Hilger's specialities. On 25 September, he provided Twyman with a list of firms to approach, suggesting there had been little headway made in organizing new sources of supply. Matters deteriorated further throughout October, and by the end of the month were so bad that Jackson travelled to London to try to resolve the problem.

There were two difficulties for Barr & Stroud. Firstly, in Jackson's words, Hilger's could not supply 'even our minimum demands for optical parts' and secondly, the War Office had asked Hilger's to restart production of the unsatisfactory Marindin rangefinder which it had bought in small numbers between 1907 and 1913.[56] To Barr & Stroud, that idea was wholly unacceptable. Many of the parts for the Marindin had actually been made in Glasgow, and Jackson knew that Hilger's had constantly struggled to maintain output and quality; its resurrection threatened to affect Barr & Stroud's output by diverting Hilger's efforts.[57] Ignoring the War Office's right to order what it chose, Jackson wrote bluntly that 'We do not see how you can undertake the manufacture', and threatened that 'if we cannot get from you the supplies on which we have been counting we shall be forced immediately to get our supplies elsewhere'. Exactly where they might come from Jackson did not say, but the threat persuaded Twyman to convince the War Office that restarting Marindin production was not feasible given the likely problems in sourcing its mechanical components.[58]

Hilger's inabilities were to some extent lessened by the Barr & Stroud's growing self-sufficiency. Jackson complained yet again to Twyman about shortcomings early in December, pointing out that Barr & Stroud had now increased its optical capacity to the point where 'we are held back by you, and only by you'.[59] In consequence, Barr & Stroud would start making the parts which Twyman had failed to deliver, and the ultimate outcome would be the exclusion of Hilger's altogether. Jackson possibly misled Twyman by implying that optical output at Glasgow had increased substantially; although new equipment had been ordered from the Standard Optical Co. in Switzerland in mid-September, there had been little enough time for it to be made, delivered and installed and operatives trained to use it.[60] Indeed, the building extensions needed to accommodate the extra plant were only started in November and could not yet have been finished.[61] It is likely that the increased optical capacity was actually Taylor, Taylor & Hobson's in Leicester, whose ability to handle large orders was becoming established, so that the need to employ subcontractors at third-hand through Hilger's was starting to diminish.

According to Barr & Stroud, Hilger's performance was indeed greatly deficient. Jackson refused to give Twyman any further orders in December, pending 'reliable information' about what improvements would be made.[62] The reply to this has not survived, but Jackson's riposte has.[63] The unfortunate Twyman was given a piece of Jackson's mind; less than 35 per cent of 1,393 items ordered had been delivered on schedule, 804 others which had not been ordered at all had been sent in error, and Twyman's claims for Hilger's level of output were 'if you excuse us saying so ... all nonsense'. Despite this, Barr & Stroud still wanted to carry on with Hilger's, and would provide weekly requirements lists so Twyman could prioritize his own deliveries. Even if alternative sources were becoming available, Hilger's were clearly still essential to maintain output. Sentiment certainly had nothing to do with it. Although well aware that costs were escalating, Jackson flatly refused to tolerate a proposed price increase, countering that he hoped Twyman would instead be able to reduce costs through 'the experience you are now gaining with manufacturing large quantities'. His dependence on Barr & Stroud left Twyman little choice but to accept both criticisms and demands. Jackson's acknowledgement of his assurances twisted the knife still further by pointing out that quality was as important as quantity and speed, and that before the war the bulk of Hilger's current output would have been rejected at Glasgow.[64]

In January 1915, Barr & Stroud began to concentrate its outside orders on Taylor, Taylor & Hobson because the Leicester firm consistently worked to high enough standards and was willing to adopt 'novel' methods in both optical and mechanical engineering to improve quality and output.[65] Jackson conceded that the Leicester-made objective lenses were better than those being made, or likely to be made at Glasgow, and asked Taylor Hobson substantially to increase production.[66] At the same time, orders placed with other suppliers began to be cancelled or not renewed. Complaints were made to both Beck and Watson, both of whom had earlier been given large orders for lenses, about late deliveries and poor quality, and ordering from both firms ended in June 1915.[67] This shifting of emphasis and reduction of dependency on a multiplicity of outside suppliers resulted principally from establishing an accommodation with a firm that was prepared to innovate to aid Barr & Stroud's output. The firm's letter books show that Jackson had never hesitated to complain about suppliers, so the lack of critical correspondence strongly supports the assumption that the working relationship with Taylor Hobson went well during the first half of 1915. Rather than simply complaining about inadequate quality, problems which arose led to collaboration in solving manufacturing difficulties, with both firms contributing equally to the effort.[68] That deliveries from Leicester met requirements for both speed and quality must have encouraged Barr & Stroud to dispense with less satisfactory suppliers, particularly if, as Jackson had claimed, the firm's own output was indeed also starting to increase. [69]

The problems of achieving adequate optical production were also eased by the plateau in Army rangefinder orders that followed those in the autumn of 1914. Unlike the other optical munitions makers, Barr & Stroud's monopoly supplier status meant that it was not faced with tendering for a multiplicity of small contracts, and during the first six months of 1915 the company could organize its production, albeit with some difficulties, in the light of a known level of demand. As a result, its output for War Office contracts kept largely on schedule, something that could not be said for most of the optical munitions industry.[70]

To increase optical output, Barr & Stroud had at first been able to employ a number of experienced makers who performed with varying degrees of success. Then it had embarked on a subcontracting exercise designed to supplement supplies without increasing its own burden of administration and coordination. When this proved unsuccessful, the firm fostered relations with a new supplier which it helped to reach a position where it could replace virtually all the other subcontractors. That there was sufficient capacity in the optical industry to be able to pick and choose before selecting one prime subcontractor suggests that the demands being placed on the industry were insufficient even to occupy its capacity let alone overwhelm it, a situation that could only have resulted from the War Office's failure to place substantial contracts across the entire cohort of optical manufacturers. In December 1914, the Periscopic Prism Co. had asked Barr & Stroud for more work, despite already getting business directly from Glasgow and via Hilger's, and was also able to take on the design of a telescopic rifle sight for the War Office. Aldis Brothers in Birmingham also had sufficient spare capacity to take subcontract work from Hilger's and to start making telescopic rifle sights of their own design.[71]

Barr & Stroud's problems were by no means confined to obtaining components from outside suppliers. Like other manufacturers, the firm was affected by the loss of men who left to 'join the colours' in response to the government's recruiting campaign, particularly those who had skills essential to the production of optical components.[72] As early as mid-August, Jackson warned the War Office that the optical workers who made up just 6 per cent of the workforce were vital to production. If they left, they could not be replaced, which would seriously compromise production.[73] Even though some of the Territorial Army members in the workforce were called up and others volunteered, there was a net increase in workers from 1,200 in July to 1,400 by late October,[74] although most of them were likely to have been unskilled or semi-skilled. The greatest labour problem in August and September 1914 turned out to be the temporary loss of 'all our best rangefinder adjusters', the highly skilled men who were 'lent' to the Navy in July to check the rangefinders of the fleet before it was dispersed to war stations.[75] Three days after the declaration, Jackson flatly refused to provide any more of them, asked for the speediest return of those in English ports, and warned that Admiralty deliveries would be 'impossible' unless the request was met.

The loss of skilled labour in the munitions industry and subsequent attempts to find replacements led to efforts in 1914 and 1915 to employ skilled Belgian mechanics who had been displaced by the German invasion.[76] In mid-December 1914, Barr & Stroud told the Board of Trade that although extra workers were urgently needed in all departments, the firm would decline to take any of the ones the Board had offered to provide.[77] The reason given was that the Admiralty had forbidden the employment of foreign workers in the factory, but this was a half-truth; what the Admiralty insisted on was that no foreign workers were allowed into the area where its contracts were being worked on. There was no reason why such workers could not have been employed elsewhere, and in fact Japanese naval personnel were already present in the factory taking part in the assembly of range-finders for the Imperial Navy.[78] Presumably the labour shortage was not critical if the company could afford to turn away skilled mechanical engineers. The com-plaint about a shortage of workers may have been a proactive device intended as a hedge against the possibility of future difficulties rather than an immediate prob-lem. What Jackson wanted was to keep men who were skilled in Barr & Stroud's methods, not to have to train and integrate foreign workers. His dealings with Hilger's illustrate a pronounced willingness to emphasize or deliberately exagger-ate the severity of a problem in order to obtain a result beneficial to the company.

Barr & Stroud and the Friction Caused by the War Office

The first year of war produced other problems that were unique to Barr & Stroud. Before the war, the firm regarded the Admiralty as its most important client, not necessarily because the Admiralty was always the largest spender, but because the Royal Navy had come to involve the company regularly in its requirements for rangefinders and continued to influence other navies by its adoption of new equip-ment. Where the Royal Navy went, others tended to follow and for Barr & Stroud the cachet of being its sole rangefinder supplier was invaluable. When European armies began to buy smaller, man-portable rangefinders in growing quantities, Barr & Stroud developed another kind of relationship with the French army as it stead-ily became by far the company's largest client. This differed from that with the

Royal Navy in being principally one of supplier and client rather than the more complex and symbiotic one of design and production interaction that existed with the Admiralty. Nevertheless, Barr & Stroud developed a sense of responsibility towards its French client because as with the Royal Navy there was the benefit of endorsement, besides the considerable income from French government business. The Admiralty had always recognized Barr & Stroud's need for overseas customers, allowing the company to balance its domestic and foreign commitments and obligations. The outbreak of war and the ascendancy of the War Office as a new major client had a significant effect on these arrangements and eventually came close to compromising Barr & Stroud's relationship with its French patron.

The possibility of conflicts of interest over new government business and existing private contracts between munitions suppliers and their customers became a matter of concern soon after the war began. The Board of Trade feared that firms might not give absolute priority to government orders when they already had existing orders to meet, as doing so might place them in breach of contract under civil law.[79] The Defence of the Realm Act already gave the Admiralty and the Army Council powers to take over part or all of the output of 'any factory or workshop in which arms, ammunition, or warlike stores' were produced, or even to take possession of the premises themselves.[80] Rangefinders were 'warlike stores' and at the outbreak of war almost all those still in Barr & Stroud's factory had been claimed by the Admiralty and the War Office. The only exceptions were French orders, for which the firm had already negotiated with the Admiralty what amounted to an immunity. However, by December, the War Office had told Barr & Stroud to limit French deliveries to no more than fifteen per week, causing an embarrassed Barr & Stroud to explain to the French Military Attaché that supplies would be delayed by several months and depended entirely on the permission of the War Office.[81] This interference with foreign business was a matter of concern to the company, not just because of its perception of responsibility to the French, but because of the overall importance of its foreign trade.

The firm's frustrations over such restrictions were evident in a letter Jackson sent in early January to the Secretary of the War Office.[82] In another indication of the efficiency of the War Office's Contracts Department, Jackson pointed out that although 'The whole of our output is at the disposal of the British Government' the War Office had actually failed to take all the rangefinders it had been offered. Those not taken were still in the factory, but they could not be sold elsewhere despite the firm still having uncompleted and urgent orders for identical instruments from Greece and France. Jackson complained of the ambiguity of War Office instructions which, if taken at face value, meant that no foreign orders at all could be dealt with until every British contract had been completely filled. He followed this with a second letter concerning a request from the United States Army to quote for rangefinders, which he had been forced under the War Office's instructions to decline.[83] With ill-disguised spleen, Jackson stressed that it had taken 'several years of trials' to get so far, and as a result 'The order for these instruments will probably be placed with an American firm [Bausch & Lomb] who are agents for a German firm [Zeiss] to the detriment of the British trade'. Having got the bit between his teeth, he wrote again the following day about a new Russian naval enquiry for thirty-two large rangefinders which, he complained, Barr & Stroud would not be able to accept because of the War Office policy, despite the fact that it 'would not affect the production of small rangefinders for the War Office'. None of these letters had any effect on the War Office; the rangefinders earmarked for Greece remained in store until 1918, and the embargo on foreign business continued in force, even after the Ministry of Munitions was created.[84]

The first nine months of the war saw the optical munitions industry make a by no means uniformly successful transition from peace to war. It was beset by problems that were rooted in the context of a war for which industrial planning was virtually non-existent and had to cope simultaneously with technical, political, economic and social problems that truly made up a plethora of variables in which the industry's artefacts had to be made and placed. In this stage the industry functioned as a number of uncoordinated and disparate units, a collection of small communities that were pushed apart rather than drawn together by the pressures of war.

This period, though, went relatively smoothly for Barr & Stroud, and the process of industrial mobilization was more successful than elsewhere in the optical munitions community. In a sense, the business was already mobilized because unlike every other, its entire product range comprised optically based systems for warfare and the lack of civil products meant that the conversion to war conditions was simpler than for anyone else in the optical industry. The most noticeable change at Glasgow was the disappearance of almost all the foreign clients and their replacement by the War Office. Matters were very much different for the rest of the optical industry which saw the gradual phasing-out of civil markets and products and their replacement by a state client whose requirements were inadequately formulated for instruments that were often unfamiliar, and whose procurement system was strange to those having to deal with it. The rest of the optical industry, or at least the part of it for which records are available, fared less well than Barr & Stroud. As a result, the output of optical munitions other than rangefinders was, by the spring of 1915, far less than needed. In this, the optical sector was no worse than, say, small arms or artillery shells, and the same remedy proposed for them by the government would be applied to instrument manufacture.[85]

The creation of the Ministry of Munitions in May 1915 was to influence the optical trades significantly, and the account now moves on to examine the mechanisms by which a large-scale and largely effective optical munitions industry was constructed out of an existing infrastructure that at first seemed reluctant to make the transition.

5 INDUSTRIAL MOBILIZATION: THE MINISTRY OF MUNITIONS AND ITS RELATIONSHIP WITH THE INDUSTRY

The second stage of the optical munitions industry's war was one of large-scale industrial mobilization, a process inextricably interwoven with the policies and attitudes of the Ministry of Munitions. Its creation in the late spring of 1915 and, in particular, the setting up of a department dedicated to optical output was responsible for increasing both the volume and diversity of production between then and the end of the war in 1918. To do so, the Ministry brought into being what can be best described as a 'conscript' optical munitions industry which largely submerged the identity of what had existed before the war and during its first ten months. The story of this mobilization is complicated by the existence within the Ministry's Optical Munitions and Glassware Department (OMGD) of parallel aims for short-term and long-term change within the whole of Britain's optical industries. The OMGD essentially looked to replace what it presented as an outmoded, inward looking, pre-war optical instruments industry with a reconstructed one that would not only meet wartime needs but be able to secure a dominant position in the foreign markets which were optimistically expected to emerge after the defeat of Germany removed its large and diverse optical industry from the international stage. These aims were sometimes in conflict with each other, and struggled to find adequate expression within a framework of problems that were grounded in shortages of materiel and an unsatisfactory technological infrastructure. The OMGD's official record of how these dimensions were managed has been largely responsible for colouring later perceptions of both pre-1915 optical manufacturing in Britain and the effectiveness of wartime measures. The story of the industry's mobilization has to look as closely into the motives and actions of the OMGD as into those of the industry in order to explain the process that followed.

Agendas and Attitudes

The creation of the OMGD marked the start of an expansion in output as well as a considerable transformation in the industry. According to the Department's own records and the later printed account in the Ministry's official history,[1] this constituted a major achievement which contributed greatly to the war effort and was attained in the face of problems within the industry that had been brought forward from peacetime. Although acknowledging the efforts made by almost all the companies it was involved with, the official account made the point that the vast majority of improvements could not have come about without OMGD's initiatives. This has been accepted by later writers who have asserted that the industry was indeed generally inadequate in education, training, organization and equipment in early 1915.[2] Some of this is true, but the deficiencies did not all apply across the whole of the industry, and parts of it had as much to teach OMGD as the department had to tell them. The story of optical munitions manufacture after mid-1915 is really of a demand-led and conscripted industry that benefited principally from the coordination and allocation of resources provided by the Ministry of Munitions, rather than being transmogrified through the direct action of the OMGD.

Even allowing for partiality in the official published record, there is no doubt that the Ministry's optical section acquired a chaotic and thoroughly unsatisfactory procurement structure whose origins have been considered in the preceding chapter. The severity of the problems in 1915 was considered to be so great that delivery prospects for optical items were then 'more unsatisfactory than in any other class of munitions' and the entire optical manufacturing industry was 'in a critical position'.[3] The condensed version of subsequent events in the Ministry's own account gives little sense of the variety and extent of the difficulties that were encountered.

A Visionary for the Industry – Frederick Cheshire

Both the shaping of the OMGD's policy and the subsequent historical perspectives of the industry owe a great deal to Frederick Cheshire (1860–1939), one of the first and subsequently the most influential of the personnel recruited by the Ministry. Cheshire was drafted in from the Patents Office by the Ministry of Munitions, initially as 'an expert on optical questions'.[4] When the OMGD was established he was appointed as its joint head with particular responsibility for what the official historian described simply as the technical side of its operations because he had 'an extensive knowledge of the [optical] trade and of the difficulties under which it worked'. This summary hardly did him justice. Professor Cheshire had entered the civil service in 1880, joined the Government Laboratory in 1882 and then transferred to the Patents Office in 1885 where he rose to become Examiner of Patents, the senior post which he held on joining the Ministry. He also continued to hold a lectureship in physics at Birkbeck College, London which had been conferred on him in 1895, and was also an Associate of the Royal College of Science and a Fellow of the Institute of Physics.[5] Cheshire's work at the Patent Office, particularly, had placed him in a privileged position to keep abreast of optical developments. There, he would have had the opportunity to see all the specifications received from British and foreign applicants, as well as the British ones suppressed from public view relating to military and naval applications of optics which the state wished to conceal in order to maintain secrecy.[6]

In 1913 he had delivered the Royal Photographic Society's prestigious annual Traill Taylor Memorial Lecture on the subject of rangefinding instruments, the text of which was subsequently published as *The Modern Rangefinder*, then one of the very few published works on the subject in the English language.[7] Cheshire was as well placed as anyone else in Britain to know the current state of optical instrument design, both for civil and military applications, and his appointment allowed him to exert a profound influence in the way energy was directed towards British optical manufacturing during the war. He quickly become the driving force in the optical department's efforts to reconstruct technical education and training in the industry in order to achieve what he saw as essential changes in its constitution, capability and performance.

Cheshire's involvement with optics was not confined to Britain. He had long-established connections with the largest and most influential firm in the extensive German optical industry, the Carl Zeiss Stiftung at Jena, which had been established in 1852. In 1902 he had co-translated Felix Auerbach's history of Zeiss's first half century and the neighbouring optical glass manufacturer Schott & Genossen which was effectively its subsidiary.[8] *The Zeiss Works* drew attention not only to the size and diverse manufacturing programmes of the two companies, but also to their emphasis on scientific training, investment in technology and attention to the education and welfare of their workforce. It was the first widely available detailed account in English of German optical engineering practice and appeared at a time when concerns over the condition of British 'opto-technics' were beginning to be voiced in England.[9] The book depicted the extent and variety of the firm's activities in considerable detail. It described the numbers and roles of staff and workers as well as a breakdown of its output and turnover, showing that no single British instrument maker was then comparable in size, vertical integration, diversity of manufacture or scale of trading. Auerbach was not a Zeiss employee but a close friend of Ernst Abbe, the principal motivating force behind the firm, and the book was really a statement of the business philosophy driving the company.[10]

To many in Britain who were involved in optical manufacturing, the book emphasized the unfavourable differences between British and German structures and practice regarding the optical trades. Although Zeiss was by no means typical of German instrument makers in general, its scale, diversity and commitment to scientific research were seen by many in Britain as the exemplar of an optical industry to be envied and emulated. That Cheshire personally was greatly impressed and influenced by the Zeiss business is beyond doubt; he referred to his 'Jena friends' in a foreword to a subsequent printing of Auerbach's book and freely acknowledged the firm's pre-eminent importance in the fields of technical optics and instrument manufacture.

Cheshire's high opinion of Zeiss was echoed by many in Britain who had come to see German optical firms and their methods as inherently superior. This sense of relative national inferiorities was well established in the industry before 1915 and had been used in attempts to procure assistance from outside the industry for its modernization. There had been unease amongst the optical instrument makers for over a decade that German companies were taking an increasingly large share of the United Kingdom's optical market.[11] Matters reached a head in 1911 when the London County Council's Education Committee, under sustained pressure from members of the largely London-based optical industry, held an enquiry into the need for organized scientific training in technical optics.[12] This concluded that the establishment of an institute for training on the lines of the German system would be highly beneficial, but because no funding was forthcoming from either the London County Council (LCC) or the industry itself, the matter was left in abeyance. The problems of munitions supply that led to the creation of the Ministry of Munitions gave those subsequently involved in the production of military optics an unprecedented opportunity to remedy what they saw as crucial shortcomings in the whole of the British optical industry.

Work on the remedial process began almost immediately after the OMGD was created. At the start of August, according to the Ministry's later account, 'the whole trade was in a critical position' regarding its deliveries of war orders.[13] Because of this, the War Office Contracts Department was said to have been unable to meet the Army's requirements for optical munitions and the optical section had to deal with a serious shortage of optical instruments. The official historian was reiterating what became OMGD orthodoxy during and after 1915: that the lack of equipment stemmed from inadequate output, and was a production rather than a procurement problem. In that view, the shortages resulted from deficiencies in organization in the manufacturing industry, inadequate equipment and a scarcity of raw materials. To make matters even worse, there was a lack of working capital which had not been properly addressed by the War Office Contracts Department in the preceding months. The situation was, by this account, at the very least, problematical. In Cheshire's contemporary and privately expressed opinion it was presented as being even worse.

On 13 August he wrote to his immediate military superior, Colonel Wedgwood, summarizing the severity of the difficulties the optical branch faced.[14] It could not have made good reading. Cheshire began by saying that 'for many years before the war broke out, the optical trade in England was a dying one'. Very few of the makers were paying dividends and he would have been surprised if any of them had been 'in a satisfactory and prosperous condition'. When the war began, the trade was in no condition to meet the demands suddenly made on it. By August 1915 output was only half of current needs, and his prognosis was that supplies would fall further behind as demands continued to grow from the Army's expansion and battlefield attrition. What Wedgwood received was a discouraging picture of a moribund industry in desperate need of assistance, not so much for its own sake but for that of the war effort. However, Cheshire must have known that his description of the pre-war optical trade by no means accorded with reality. His assertion that only 'a few firms had been more or less kept alive' was, to say the least, misleading. Leaving aside those already involved in optical munitions manufacture in 1914 and mentioned in the preceding chapter, there were at least fifteen more optical instrument makers who were very much in business producing optical instruments.[15] Whatever his reasons for writing in such a vein, that it was no temporary aberration is shown by a later draft report written on the progress made up to October 1917.[16] The report stated that when the war began, the country had been in a 'deplorable condition' regarding its ability to produce optical munitions on a large scale, and that the machinery employed within the optical trade was inadequate and antiquated. Warming to its theme, it went to say that 'The workshops were shanties', and the trade as a whole, in the opinion of 'many men in a position to judge, "already dead and damned"'.[17] Given Cheshire's experience and knowledge, such remarks must raise the question of motive. Why should he project such a misleading image in the first place, and why maintain it two years later?

His depiction of the industry's condition in 1915 may have been highly coloured, but his concerns were nevertheless genuine. They were grounded less in the fear that the manufacturing base was then woefully inadequate to supply the armed forces' immediate needs than that civil optical manufacturing in Britain would sooner rather than later be completely outclassed by Germany's. With the exception of Barr & Stroud, all the peacetime optical munitions makers had relied on civilian trade to provide most of their incomes. Cheshire realized that such commercial work would be vital to sustain a substantial post-war optical munitions capacity and that if businesses were to become unprofitable, failures in the optical industry could eliminate much, or possibly even all, of the capacity to manufacture for the nation's defence. The 1911 LCC Report had highlighted apparently serious flaws in the structure of the British optical industry, the strongest evidence for which was that the country was a net importer of optical instruments.[18] That conclusion was derived from data published in the *Annual Statement of the Trade of the United Kingdom* in the period up to 1909, but it was not in fact properly justified.[19]

The LCC Report acknowledged that the classification providing the data – 'Scientific Instruments and Apparatus other than Electrical' – lumped together optical instruments with other items, such as photographic film and printing paper, which were not actually part of the optical sector, making it difficult to identify and assess accurately the size of the optical component. Despite that problem, the report's authors were content to assume that because the whole category was in deficit to the tune of £157,000 in 1909, then so must be optical goods. In 1910, however, Customs and Excise removed most of the non-optical items from the category, to leave behind telescopes of all kinds, photographic cameras and their lenses, microscopes and ophthalmic apparatus, lenses and prisms for scientific instruments and surveying instruments.[20] This rearrangement caused a wholly different picture to emerge (see Table 5.1) although no attention was subsequently drawn to it, least of all by Cheshire, for whom it would have tended to weaken the case for reforming the industry which he was starting to construct.

Table 5.1: Balance of trade of scientific instruments, 1911–14 (amounts shown in sterling)[21]

Year	Imports £	Exports £	Surplus £
1911	555,106	713,328	158,222
1912	645,379	707,061	61,682
1913	710,341	767,402	57,061
1914*	471,525	646,493	174,968

* The outbreak of war curtailed both imports and exports.

Even though this showed a declining trend up to 1913, the optical trade in toto was clearly in a far from a terminal condition, even using the yardstick of overseas trade as the sole arbiter. The underlying condition was really one of relative rather than absolute decline, a condition also identified in other industries.[22] The total demand for optical goods was expanding and although imports were rising, so were exports. The level of domestic optical production was actually increasing, even without the figures for naval and military rangefinders which, as 'munitions of war', were not counted with civil trade in the *Annual Statement*. Had the munitions component been included, the recorded surplus would have been even larger. Taking into account the known data for the War Office's imports of all types of optical munitions, and Barr & Stroud's export orders for rangefinders, a far less unfavourable picture emerges (see Table 5.2 on p. 113):

Table 5.2: Balance of trade of scientific instruments and optical munitions combined, 1911–14 (amounts shown in sterling)[23]

Year	Civil surplus £	Rangefinder exports £	War Office imports £	Total £
1911	158,222	50,241	-7,428	201,035
1912	61,682	160,768	-10,579	211,871
1913	57,061	185,330	-4,631	237,760
1914	174,968	114,057	nil	289,025

These figures also indicate the substantial proportion of optical munitions exports relative to those of the whole optical instruments industry. In 1912 they accounted for 22.7 per cent of the total, and 24.2 per cent in 1913. The declaration of war disturbed trading patterns and curtailed exports of optical munitions so that a meaningful picture for 1914 is hard to arrive at.

But Cheshire was not concerned in June 1915 to point out hidden strengths in the greater optical industry. To the contrary, he wanted first to highlight and then to rectify the perceived weaknesses in finance, scientific education and training, as well as the slow adaptation of modern technologies, that were seen as debilitating influences. To go about this, he deliberately emphasized the existing idea that the optical industry was inadequate for the task it now faced and that the output of urgently needed devices would inevitably be compromised. Whether or not that notion was entirely justified, the scale of the immediate problem certainly demanded some sort of solution that would quickly ameliorate the undeniably difficult situation he had to manage. What Cheshire faced was the simultaneous need to overcome what he was representing as a critical situation, as well as long-term measures to lift the industry out of the circumstances that were alleged to have brought about the predicament. His confidence in being able to tackle the problem is shown by the speed with which he defined the optical section's role. By the time he first wrote to Wedgwood he had already identified the work needed as falling under five headings:[24]

- to provide financial aid where necessary
- to supply expert technical advice
- to expedite deliveries of raw materials and components
- to provide trained labour
- to establish research centres 'to set the trade on a sound basis'.

The first three items on the list might be implemented quickly, but the fourth and fifth would certainly require more time to bring about. The aim of setting up research centres was in harmony with the long-term goal of the progressives in the industry since 1902, which was to provide advanced scientific training and establish a tradition of systematic technological research on the lines already being followed in Germany with the involvement of technical training schools and universities.[25]

Whatever the long-term goals Cheshire harboured, there were major problems to be addressed urgently before the OMGD could begin to build a coherent and efficient optical munitions industry out of the chaotic conditions in the summer of 1915. First, those firms already involved in military production lacked any real motivation to increase output because of previous War Office contracting policies. Then, making matters worse, many businesses felt threatened by the prospect of control under the Munitions of War Act which been passed on 2 July. Besides those questions of corporate morale there was a shortage of machine tooling and skilled labour, as well as a serious scarcity of the special glasses needed to make lens and prism systems. And finally, there was no trade organization within the optical industry that might facilitate coordination between its constituents. Despite the sweeping powers conferred on the Ministry, and Cheshire's optimism, the OMGD frequently found it hard to bring about the changes it regarded as essential to the industry's improvement.

The Application of Pragmatics

To apply Cheshire's five-point action plan, the OMGD was divided into technical and administrative sections. Although previous writers have concentrated on the efforts and achievements of Frederick Cheshire and the technical section, so far as the contemporary industry was concerned the head of the administrative side had a more immediately important role to play. Little is known about the background of the man appointed to that task, Alfred Esslemont, beyond his being a Fellow of the Optical Society.[26]. His post in the Ministry suggests he was engaged for his combination of organizational abilities and technical knowledge, and throughout his work with the OMGD he was constantly engaged in liaison with both the instruments and glass industries, becoming a Director of the Department in 1917. Esslemont not only had to create a departmental structure that could bring some sort of order from the chaos of 1914 and early 1915, he also had to persuade the trade to adopt new working methods and to accept the subsequent imposition of the state's wartime controls. The administrative side of the OMGD extended far beyond keeping records and allocating contracts; it overlapped the remit of Cheshire's department, spreading into technical matters such as instrument and machinery design. Most importantly, it came to assume a diplomatic role between industry and the state. The importance of the administrative section should not be underestimated; particularly under Esslemont, who died in 1918 before the end of the war, it was vital to the success of the wartime industry.

Esslemont's first task was to motivate the existing War Office contractors to increase their output, and in June 1915 he called them together to address 'the dangerous condition of affairs'.[27] According to the OMGD, the previous dealings of the War Office Contracts Department with the trade had been so bad as to create a 'most paralyzing effect' on it, and firms had generally become dissatisfied with government contracting, strongly distrusting the methods used in placing orders. Their chief complaint, in an echo of the Boer War, was that no company ever received an order large enough to justify setting up for quantity production, with contracts for small numbers of instruments being scattered amongst a number of firms.[28] In short, they found it hard to make money working for the War Office. Its ordering policy, founded on considerations of peacetime fiscal probity and perpetuated by institutional inertia, had created a delivery situation that was unsatisfactory for everyone. Esslemont managed to reassure the manufacturers that, under the OMGD's umbrella, they would now be dealt with as soon as possible with regard to the distribution of orders. It was not only the instrument makers that Esslemont sought to mollify and galvanize, he also needed urgently to take up the question of optical glass.

Although not strictly part of the optical munitions industry, a diversion into the production of the special glasses required for the manufacture of optical instruments cannot be avoided and this is probably a good place to make it. Optical glass has been mentioned on several occasions as the story has progressed, but it must be emphasized how important its provision was during the war.

Optical glass is quite unlike the material used for windows, bottles, test tubes or petri dishes. Lenses and prisms for optical instruments require glasses which are entirely free from discolouration, physical blemishes and inherent stresses which would distort the image being created by the instrument in which it is used. Optical glasses have particular qualities of refraction and dispersion which influence how efficiently and to what degree the lenses and prisms in an optical system bend and treat the light rays passing through it. Different combinations of those characteristics are used to achieve specific goals by designers of instruments, often requiring several types of glass in even a relatively simple instrument such as a telescope. Without such glass, instrument making simply cannot be done.

Its manufacture was, and largely remains, a slow and difficult process which in 1914 had been done only on a small scale in Britain.[29] Most lens makers preferred to use German or French glasses because of a perception that their quality was superior. The only domestic maker was Chance Brothers of Birmingham whose main business was making a wide range of commercial and industrial glasses.

Prior to August 1914, some 70 per cent of British requirements for optical munitions glass had reputedly come from Germany, a quarter from France and the balance of 5 per cent was produced domestically.[30] German imports had ceased almost immediately when war began, according to Chance, and increasing demand rapidly depleted British stocks of both the more common and sophisticated glasses, leading to early shortages of suitable materials.[31] To make up the deficiency, the War Office began negotiations with Chance in late 1914 to increase production and arrangements were made to obtain deliveries from the only substantial producer of French glass, Parra Mantois.[32] By the time the Ministry of Munitions was created the War Office had still to come to any agreement with Chance Brothers about obtaining greater output. Making a difficult situation even worse, supplies from France were being withheld because the War Office had disrupted an agreement whereby optical glass was delivered as a quid pro quo for the supply of Barr & Stroud rangefinders.[33] The glass situation was, to say the very least, demanding urgent attention.

To try to conclude the War Office's unsuccessful negotiations Esslemont began discussions with Chance Bros in June 1915. The firm maintained that its continuing inability to increase output was caused by the lack of technical staff and insufficient capacity for expansion. Its optical glass making was only a small part of a much larger business largely producing glasses for more mundane uses such as bottles and windows. The specialized types needed for lenses and prisms had previously made little profit and, to some extent, their manufacture had been continued on patriotic grounds.[34] The firm had no pool of scientists for researching new formulations and was disinclined to embark on that route again because their previous unacceptable financial losses made them unwilling to commit themselves to investment which might have no future return.[35] Another argument, which might have carried weight in late 1914 but certainly not in mid-1915, was that an early end to the war would mean the familiar German glasses again being freely available to firms which were used to working with them. Esslemont's tactic was one of stick and carrot. The nation desperately needed what only Chance was able but unwilling to supply; in exchange for a massive patriotic effort on the company's part, the Ministry of Munitions would guarantee it a ten-year monopoly of government business for optical glass. That was indeed a tempting carrot, but the company had to agree to provide plant and staff sufficient for a monthly output of 8,500 lbs of optical glass throughout the war, plus maintaining plant and a glass reserve in peacetime, in exchange for its monopoly.[36] The promise that 'guarantees', i.e. loans, grants and the availability of scientifically qualified staff, would be provided to assist the meeting of Esslemont's ambitious production targets re-wakened Chance's dormant spirit of patriotism and its expansion programme went ahead even though the details were not finalized until January 1916.[37]

Esslemont's rejuvenation of Chance Brothers in 1915 opened the way for a larger effort to overhaul optical and scientific glass manufacture in Britain, which by the end of the war had gone a long way to making the country independent of foreign supplies, but in this story his critical role was in overcoming another impasse inherited from the first months of the war.

But to return to the optical munitions makers themselves; after the general meeting in June 1915, a subsequent series of meetings with individual companies, along with the offer of some financial assistance were, perhaps optimistically, thought to be enough to ensure that 'a maximum of effort' would now be made, and as evidence of this the OMGD noted in late July that some makers were spending significant amounts of their own money on their own initiative. Output then began to increase 'by leaps and bounds' according to the later words of the Department.[38] Its own recording system did indeed register a substantial increase during late 1915 but, as not infrequently happened with the OMGD, matters were not quite what they appeared to be.

Its figures call for some comment. July's output, before the department could have had any effect, was just over £9,000 for all optical contractors employed on War Office business. By October that had risen to nearly £16,600 – an increase of almost 81 per cent on July. This may have looked encouraging, but as a proportion of the industry's total capacity it was actually miniscule. Gross exports alone of civil 'scientific instruments' in 1913 had been £767,402. October's output amounted to an annual total of scarcely £200,000, which is so greatly removed from 1913's levels that the OMGD's figures need treating with some circumspection. It is unlikely that commercial production still accounted for the largest part of output because the Defence of the Realm Amendment Act had already empowered the Admiralty and the War Office to obtain precedence for their orders by requiring makers to put aside other work,[39] and by mid-1915 it is extremely doubtful if any makers had not been substantially affected by the changed circumstances of the war. Even by June 1916 the monthly total was still under £40,000 and at the year's end had scarcely reached £47,000, still far less than the average in 1913. One explanation for the low figures is that they referred to the value of items held after acceptance inspection at Woolwich Arsenal, a process which, as will be related later, continually created backlogs; only after instruments' satisfactory inspection were invoices passed for payment and then recorded by the OMGD. Another possibility was that instruments being made for the state were of lesser value than civil ones. A third explanation, and one that would have best fitted Cheshire's agenda, was that the industry's output had indeed collapsed and was incapable of improvement without substantial assistance.

Irrespective of whether the figures being generated were inadequate or misleading, Esslemont's reports of his June meetings indicated that problems in War Office procurement procedures had been as much to blame as any structural failings in the industry, although this was not something that was stressed either in correspondence with the senior Ministry personnel or later in the official history. One way he proposed to reassure and encourage the trade was by making the distribution of contracts a matter for the OMGD itself.[40] Dealing with an agency whose managers understood their problems would doubtless have been a vast improvement for the makers, but the OMGD had to argue strongly to justify this departure from what was intended to be standard Ministry practice. Cheshire and Esslemont boldly asserted that optical munitions orders were more complex than others dealt with by the Ministry and were 'enormously complicated' by shortages of labour and materials as well as changes to specifications. Almost every contract required 'detailed and expert knowledge' by the staff responsible for its placement and subsequent oversight.[41] The suggestion was that the optical industry was so highly specialized that only dedicated, technically competent staff within the Ministry would be able to manage dealings with it. This was not really accurate, but Cheshire and Esslemont had quickly produced such an image of failure and chaos that it allowed them to employ an argument that was actually intended to let them direct and manage the nascent optical munitions industry's development in order to maximize the benefits of the civil optical instruments sector which would derive from the expansion of wartime business. Their lobbying succeeded, and for the rest of the war the OMGD continued, uniquely in the Ministry, to operate its contracts directly with the trade.

A substantial part of Esslemont's success in gaining assurances of cooperation from the makers resulted from assuring them that orders would now be placed on a large scale and for a considerable length of time. His ability to do that came from the political recognition that the war would be measured in years, with the accompanying necessity to ensure a continual, large-scale supply of munitions. That understanding had been responsible for the creation of the Ministry of Munitions, whose powers to direct and control industry were vested in the Munitions of War Act which came into force on 2 July 1915. Although this legislation gave the OMGD considerable powers to assist the optical industry, some of its provisions quickly created considerable unease amongst the recently mollified manufacturers and led to friction between the optical department and its political masters at the Ministry.

Control and Profits

Industry in general had been subject to controls even before the Munitions Act was passed. In March 1915, the Defence of the Realm (Amendment) Act (DORA) allowed the armed services to require manufacturers 'to give precedence to the completion of all orders and contracts' for government work, to ensure that neither commercial nor foreign government business obstructed production for the state.[42] DORA's Regulation 30A prohibited 'all dealing in optical instruments which are of service to the Admiralty and War Office ... except under special permit'.[43] This was intended to, and according to the OMGD actually did, bring about 'the extinction of private work' in the optical trade.[44] Even before the 1915 amendment, for instance, Barr & Stroud had told the London instrument makers Negretti & Zambra it could no longer sell rangefinders to individual Army officers, and informed Dollond & Co. that it would be 'many months' before there could be any hope of supplying any non-government clients.[45] Such constraints may have been irksome, but they were nothing compared to those embodied in the Munitions Act itself.

Its purpose was to further 'the efficient manufacture ... and supply of munitions for the present war', by imposing a body of regulation on both employers and workers.[46] For the optical munitions industry there were serious implications in the creation of what was to be known as the 'controlled establishment' and the attendant regulation of profits. Broadly speaking, the Act sought to remove restrictive practices on the part of labour and as a quid pro quo to limit the profits employers might make from war work. The legislation gave the Minister of Munitions the power to declare as a 'controlled establishment' any business engaged on munitions production and whose output was considered as essential for 'the successful prosecution of the war'.[47] Any firm involved in optical contracting for the Services was therefore likely to be placed under Ministry control. The limitations on profits were seen by both Cheshire and Esslemont as an obstacle to improvements in the industry which they attempted to insulate from direct control in much the same way that they had done with the placing and management of contracts.

Control under the Act was not meant to 'involve any interference ... with the management of the firm'.[48] Rather, it was supposed to relieve companies of 'the restraints imposed by trade union restrictions'. Useful as that might have been, it also greatly restricted the profits to be gained from government contracting. It was this issue that most worried the optical makers who, having just been assured that they might expect a large volume of business for some considerable time, now faced the prospect of having their future earnings greatly diminished. The Act by no means prohibited the making of profits; instead it incorporated a formula restricting the proportion of them that a business could retain. A 'standard amount of profit' for a firm was to be determined by averaging its pre-tax profits for its last two financial years before August 1914. All but 20 per cent of profits in excess of that would now go to the Treasury, the remainder being taxed at the same rate as the standard profit. Although this was seen as politically essential to maintain the cooperation of the trade unions, it was also a decided disincentive for businesses to maximize output, particularly where increased investment was needed to handle enlarged volumes of war work. The legislation also demanded that contractors did not hinder production in any way, which meant that a firm coming under control stood to be locked into meeting the Ministry's demands at levels of profit that were significantly restrained. What made matters worse for the optical trade was that, according to Cheshire, its profits before the war had been far from good.[49] In mid-August he told the Ministry that very few companies had been paying dividends at all and 'indeed, it would be surprising to learn that a single one of these important firms was in a satisfactory and prosperous condition'. In his opinion, the placing of existing contractors under control would disadvantage them financially compared to firms left outside. Esslemont echoed this sentiment, which had already been expressed by some of the optical companies, adding that attempts to increase output depended largely on their 'goodwill' and so it was important 'to take notice of their points of view' if cooperation was to be maintained. However, despite the firmness of Cheshire's assertions, the accuracy of his profit assessments is open to doubt.

Relatively few financial records survive for the firms Cheshire was writing about, but some have left figures that allow some evaluation of his comments.[50] Barr & Stroud's audited accounts show pre-tax profits for 1912 and 1913 as £32,555 and £59,530 respectively, on turnovers of £126,593 and £188,007, representing net margins of 25.7 per cent and 31.7 per cent and returns of 16.3 per cent and 29.8 per cent on the share capital employed. These are hardly those of a chronically ailing company. Thomas Cooke & Sons Ltd of York for the same years had profits of £7,287 and £7,899, with dividends of £2,615 and £5,545 to be paid from them. Taylor, Taylor & Hobson of Leicester recorded pre-tax profits of £2,433 in the trading year for 1912 after payment of unspecified preferential dividends and new buildings costing £3,663. For 1913 the figure was £8,130 after dividends and a further factory extension costing £6,400. Although the sales figures for the second two firms are not recorded, both were much smaller businesses than Barr & Stroud and their percentage margins seem to have caused no concerns in their records. These three companies, at least, certainly did not fit the image painted by Cheshire.

The accounts of the London company Troughton & Simms, however, give a picture which is less good. The firm was similar in size to Cooke's and had produced a similar product range, although it had little or no previous background in munitions production.[51] In 1912 there was a pre-tax profit of £4,347 on sales of £26,292, but in 1913 only £1,169 on turnover of £21,921, with no indication of any substantial expenditure on plant or premises to account for the reduction.[52] Not only were the end results poorer than the others', but the business apparently ran inefficiently, at least in comparison to Barr & Stroud. Profit on turnover for those years was only 16.5 per cent and 5.3 per cent, compared to 25.7 per cent and 31.7 per cent at Barr & Stroud, while 'stock in trade' was equal to 47 per cent of 1913's sales against Barr & Stroud's 12.4 per cent. Wage costs were over 57 per cent of sales, a proportion more than half as much again as the larger firm. Despite all this, 'less good' is by no means the same as 'bad', and Troughton & Simms was certainly solvent in 1913 with £11,800 cash in its bank, current trade debtors owing £3,884 and trade creditors standing at only £759.

Cheshire's woebegone depiction of the industry's financial condition was substantially at odds with these companies. To what extent he was aware of trade's detailed economic circumstances is uncertain, but his connections with it must have given him some indication of its general state. This may indeed have been less than satisfactory, but the fact that none of these businesses were in financial difficulties – with most actually solidly profitable – in 1912 and 1913 implies that things were by no means generally as bad as he alleged. The wording of his minute to Wedgwood suggests he was leaving himself some latitude in what he was saying. Terms such as 'prosperous' and 'satisfactory' are hardly precise and could be applied to other areas besides financial performance, so Cheshire may have had in mind additional, less easily quantified aspects of business performance such as scientific and technological expertise, a theme to which both he and Esslemont would later return.

Keeping the optical companies outside control in 1915 would have given them the opportunity to benefit from a growth in business that stood to generate profits very much greater than in peacetime or the first year of the war, potentially aiding Cheshire's intention to advance the scientific and technological basis of the optical industry. Long-term improvements would require companies to be financially sound and although state loans might overcome immediate cash shortages they would ultimately need repaying out of future profits. Expansion of output would require large spending on new machinery and premises and it was by no means clear in the early form of the Munitions Act that the state would cover any portion of this expenditure or permit businesses to retain any significant financial benefit from it. Even though the Act provided for the introduction of rules to allow for 'any special circumstances such as increase of output, provision of new machinery or plant, alteration of capital or other matters which require special consideration' in assessing profits, none had yet been formulated when Cheshire wrote his August minute about excluding the optical industry from control.[53] They only appeared in September, 'after long and exhaustive discussion within the Department' and would have gone a long way to meeting Cheshire's aims, but before any firm arrangements were made for treating special cases, matters were interrupted by a Finance Bill which introduced the idea of an Excess Profits Duty (EPD) to tax at a higher than normal level all profits 'in excess of a pre-war standard'.[54]

This proposal had serious implications for Cheshire's desire to retain profits within the industry. It intended to take 50 per cent of all 'excess' earnings from the outbreak of war, and unlike the Munitions Act allowed only for 'exceptional earnings and redundancy of plant'.[55] The Ministry of Munitions recognized that the EPD proposals were likely to overlap the munitions levy and saw the prospect of controlled firms being liable to pay both as an illogical and unreasonable situation. Negotiations between the Ministry and the Treasury to exempt controlled firms from the new proposals were inconclusive, and Cheshire and Esslemont continued to lobby persistently and tenaciously to keep the optical contractors outside control.

Both measures stood to deny the optical industry the opportunity for the earnings of which the previous lack was, according to Cheshire, responsible for much of its woes. Faced with still-incomplete rules in the Munitions Act and even greater uncertainty over the proposed finance bill, Cheshire's only feasible strategy was to plead that the optical firms' inclusion would result in the loss of cooperation and a consequent catastrophic decline in output. Much of his denigratory comments on the industry in the confidential internal minutes can only be interpreted as deliberate hyperbole intended to increase the strength of his pleadings whilst concealing the true reason for them, which would have been wholly unacceptable in the political context of the Munitions Act. The Ministry of Munitions was, as the official *History* subsequently reminded its readers, first of all concerned with the output of munitions, not with the profits of business in general,[56] but Cheshire's tactics indicate that he was as much concerned with earnings for the industry as with output for the state.

It has been pointed out elsewhere that the OMGD dealt with its industry in ways different to other sections of the Ministry of Munitions, but the important underlying reason has not previously been identified.[57] Instead, the suggestion has been that Esslemont opposed control simply because 'some instrument firms were reluctant' to be controlled and that their wishes had to be considered, whilst Cheshire wished to avoid what he described as 'upsetting a very delicate balance' between the trade and his department. The argument he used was that firms were accustomed to being in charge of their own affairs and were stimulated by the need to create profits but, under the provisions of the Munitions Bill they would lose their control and not be guaranteed any profit unless they had been showing a substantial one before the war. This could only have been a deliberate distortion of the truth; Cheshire must have known that the Munitions Act did not interfere with routine management and there was provision in it for dealing with previously unprofitable businesses. The reality was that the joint directors of the OMGD were deliberately seeking to pursue a policy that ran counter to the very heart of the principles of the Munitions Act in order to let contractors derive substantial extra financial benefits from their war work in order to benefit the entire industry's efficiency and competitiveness once the war ended.

These efforts enjoyed some success. In July, Barr & Stroud, Cooke's, Heath & Co., Ross, Troughton & Simms and Carl Zeiss (London) Ltd had been placed on the list of firms to be controlled, but all were quickly (if only temporarily) removed, despite reservations and opposition within the Ministry.[58] In late July the Ministry line was that firms should be controlled irrespective of their inclinations and that 'it would be difficult to leave them off the controlled lists'.[59] Opinion then further hardened in favour of listing, control being depicted as having real advantages, benefits to which only one firm, Ross, had so far objected. On 9 August a decision was taken to go ahead, but by 14 August Cheshire had persuaded Wedgwood and Eric Geddes to tell the Director General of Munitions Supply that the Ministry now felt the need for a strong case to justify putting them under control, a complete reversal of the earlier position. The Director General, F. W. Black, conceded the point although he excepted Barr & Stroud because of its engineering content.[60] Cheshire's pleadings, perhaps better described as lobbying, on behalf of the optical instruments makers had been, for the moment, successful in their guise of creating an efficient optical munitions industry.

The only other optical firms taken under control in the remainder of 1915 were Ross and Carl Zeiss (London) Ltd, both at their own request. Ross's abrupt change of heart may have been because the Ministry was prepared to put up a sizeable proportion of the £25,000 the firm had earlier decided to spend enlarging its works to handle an order for 2,000 artillery dial sights worth over £110,000 which the OMGD had placed in August.[61] Ross had previously suggested that if placed under control, the firm would be unwilling to spend its own money on the project, so the request to be designated a controlled establishment was likely to have been more of a quid pro quo than anything else. Carl Zeiss (London) Ltd was a small subsidiary of the very large German Zeiss organization which had been set up in 1909. Since the outbreak of war it had been operating in a kind of limbo, cut off from its parent, unsure of its future and, ironically, assembling binoculars for the War Office from what were probably dwindling stocks of pre-war German-made components.[62] Matters came to a head when its directors and senior managers were interned in 1915, and the business put in the hands of a controller from the Board of Trade who doubtless found that control represented a solution to its current problems as well as removing any stigma that attached to it from being an 'enemy' firm.

In January 1916, political pressure caused the Ministry to reconsider the status of the optical firms still outside control. Cheshire's success in presenting the optical industry as a special case was beginning to rebound, the comment being made that the exclusion of an industry of such importance to the war effort left its Minister, Lloyd George, with no defence to any objection to the decision. On 1 February, Esslemont was reminded that he had still to provide the names of firms to be placed on the controlled list but he continued to prevaricate, replying through OMGD's liaison officer that the Department could see 'no justifiable reason' to include any other optical companies. In his view neither the industry nor the Ministry would derive any benefit from control.[63]

The tenacity with which Cheshire and Esslemont sought to keep most of the industry outside control reflects how much they were concerned that it would hinder not so much the creation of a temporary wartime munitions industry, but the long-term growth of the entire optical industry, which they saw as being the future locus of optical munitions manufacture. Unfortunately for them, the political implications of their tactics were too serious to allow success. Wedgwood's letter expressing Esslemont's ideas brought a measured and perhaps slightly menacing riposte from the Ministry's Owen H. Smith, who wrote that not only were the continued exclusions exposing the Minister to strong criticism but continuing them could have unwelcome consequences. 'If there is a real reason for exclusion, can you send me a short minute?' he ended, suggesting he had recognized that the arguments being put forward were at best contrived. Wedgwood's reply had to admit there was little to add beyond reiterating that only the delicate treatment of the trade had permitted improved output, and that he was concerned about the effects of anything disturbing current circumstances. Smith was unmoved and on 1 March told the Director General of Munitions Supplies that he considered fears about reduced output unjustified and that there were no grounds for exempting an industry of such critical importance in the munitions sector.[64] That marked the end of attempts to keep the optical trade outside the financial constraints of the Munitions Act, and Cheshire and Esslemont's energies were henceforth more closely focused on other efforts to transform the industry.

Scientific Training and Technical Education

For Cheshire, the other pillar of support for the optical industry was significantly better scientific and technical education. In late 1915 and 1916 the case for introducing new centres to provide this was reinforced by the argument that the existing industry was constantly being held back from meeting its production targets because there were not enough 'skilled workmen, designers and [lens] computers available'.[65] This was substantially correct, although other factors such as the lack of machinery, tooling and factory space were greater and more immediate impediments to increasing output of equipment already in service.[66] The LCC's Education Department took up the work it had done in 1911 and in June 1916 its Education Officer, Robert Blair, suggested a 'national scheme for training in technical optics'.[67] The only facility that then existed was at the Northampton Polytechnic Institute at Clerkenwell in London, the location of much of the London optical trade, which provided craft training for the production of optical components, rather than higher education for design and computation.[68] Despite the immediate and pressing need to increase output, Blair's consultations with the optical industry and the Ministry of Munitions had persuaded him that the manufacturers were 'less preoccupied with the need to train workmen, than with the need to secure advanced postgraduate and research work', and his recommendations took that requirement principally into account.[69] To address the output problem, the existing Clerkenwell scheme would be enlarged to take sixty students at a time and evening classes would be provided at what were to be called junior technical schools, but most importantly a new department would be created at Imperial College to provide undergraduate and postgraduate training, as well as facilities for research workers. That would provide a centre of excellence for optics which, although familiar in Germany, was still unknown in Britain.

Blair's suggestions were adopted and in May the following year Frederick Cheshire was appointed as Professor of Technical Optics at Imperial College. This meant his stepping down as joint head of OMGD, but he maintained a connection through reverting to his original Ministry post of 'expert advisor in technical optics'. The extent to which Cheshire influenced Blair's thinking must have been considerable, particularly as the greatest energies were devoted to the area where the immediate demand was actually least urgent. Despite the assertion that output was threatened by a lack of designers, the inescapable truth remains that in 1915 and 1916 what the War Office chiefly wanted was more of the instruments that were already in production such as rangefinders, dial sights, prism binoculars and telescopes.[70] Where novel items were required, such as the telescopic rifle sight, the design was able to be done by existing firms who already had competent optical designers.[71] What the wartime munitions industry immediately needed most was extra capacity, but the OMGD, heavily influenced by Cheshire, was convinced that in the context of the long term, what the optical instruments industry needed was actually quite different.

Once Blair's ideas were accepted by the LCC, the Department of Scientific and Industrial Research and Imperial College, an advisory committee was set up to coordinate progress. Although the OMGD had no official part in what Blair proposed, it was represented on this Technical Optics Committee by Alfred Esslemont, along with members of the 'optical trades', namely Frank Twyman from Adam Hilger Ltd, Conrad Beck from R. & J. Beck Ltd and T. Watson-Baker from W. Watson & Sons Ltd.[72] Twyman, although now totally (if temporarily) committed to munitions contracting, was principally interested in scientific instrument manufacture, as were Beck and Watson-Baker. Before the war, although all had been engaged in optical munitions manufacture, their businesses had not been as heavily involved as others; Ottway and Ross both did more government contracting, and Barr & Stroud was totally committed to military and naval work. Beck, Twyman and Watson-Baker had been anxious to advance the instruments industry before the war, and Beck in particular had previously been involved in efforts to establish a makers' association, which had continued even after the war began. Their concept of what optical production would most benefit from was almost certainly more closely in tune with Cheshire's than anyone else in the London industry. Such influence was also reflected in the other appointments to the new department at Imperial College, with two of Watson's designers filling key posts.

Just before the first classes at Imperial College started in 1917, Cheshire published an open letter to the trade in the journal *Optician* that amounted to a summary of his policy for reforming optical manufacture in Britain, justifying it by the experiences of the war.[73] It was, he said, only the pressure of wartime demands that had impressed on the makers the need for scientific method to replace the old ways of trial and error, and had provided a climate where changes would be accepted. The setting up of the department at Imperial College was the start of a systematic approach to optical design that, by implication, was long overdue. There is no doubt that much of what Cheshire wrote was true, but it did not apply universally. In particular, the three firms represented on the advisory committee were no strangers to scientific method, nor were companies such as Barr & Stroud, Ross and Taylor, Taylor & Hobson. What the facilities at Imperial College were meant to do was to provide a pool of scientifically trained opticians that would, it was hoped, bolster the abilities not only of companies who already employed such staff, but also of those who had previously been unable to hire skilled designers. In June 1918, the Technical Optics Committee summarized Imperial's first optical academic year by saying the courses would meet pressing demand for highly competent lens designers and optical system computers and would eventually be able to fill the senior posts in the industry.

Cheshire's policies for long-term improvements became more relevant to optical munitions production after late 1916, and his enthusiasm for using short-term pressures to promote progress on a broader front should not be allowed to distract from the benefits which parts of the munitions industry gained from his work. Whether, in the context of immediate wartime demand, he actually produced the most appropriate short-term solution for the industry as a whole is open to debate. The War Office had never prepared any optical specifications in detail, nor had it recorded the optical designs of equipment in service, so businesses coming into optical munitions manufacture were faced with designing lens systems for themselves, even when they were making an instrument of a pattern already in use. Once the OMGD began to look for new sources of supply in late 1915 and 1916, the problems inherent in this situation quickly made themselves apparent.

For the well-established firms like Beck and Ross which had trained staff experienced in optical computing, working out lens systems for instruments like prismatic binoculars and sighting telescopes was no problem, although the work was extremely time consuming in an era when all calculations had to be done using logarithmic tables and slide rules. But because such companies were fully occupied and there were no detailed War Office specifications for manufacturing any optical systems, the OMGD was increasingly relying on newcomers to optical munitions work to do such calculations for themselves. However, in 1916 there was a chronic shortage of qualified computing personnel able to tackle the calculations needed for even the simplest of optical systems which stood to seriously impede bringing extra manufacturing capacity into being. Cheshire's planned three-year courses were meant to provide training that went far beyond what was now urgently needed, and so he and the chief designer of Watson & Sons, Eugen Conrady, set up six-week 'crash courses' in basic lens computation at Imperial College in late 1916.[74] Their purpose was to provide sufficient skills to allow the preparation of prism binocular and terrestrial-telescope systems for gun sights by people with no previous higher mathematical training. Such training had variable degrees of success, which is perhaps why they received no mention in the Ministry's published history. In one case, the course proved adequate to allow in-house design at Kershaw's entirely new factory set up in Leeds for binocular production in 1917 without needing to rely on any outside aid.[75]. However, they experienced problems actually making their own design and in March the following year changed over to making a system designed by Taylor, Taylor & Hobson which was simpler to manufacture.[76] In February 1918, the Sherwood Optical Co., a firm which was originally set up in 1915 and only recently given binocular contracts, was reported by the OMGD's Technical Branch to have suspended its output as it was waiting for a 'new optical system to be calculated by Mr Chalmers' to replace the one originally employed.[77] In Kershaw's case, the problem was rooted in the designer's relative inexperience with design and the resulting inappropriate choice of optical glass, emphasizing the intrinsic shortcomings in the short, intensive training programme.

The Problems of Failure to Standardize Designs

One key reason for the introduction of the 'crash courses' was the absence of properly standardized specifications, even for designs which were in use before the war. Despite the benefits of such standardization in facilitating production across a number of makers, the OMGD seemingly never seriously tackled the lack, despite the problems which it regularly caused in extending sources of supply. Unlike small arms, for instance, where both component dimensions and materials were strictly specified to ensure all parts were interchangeable irrespective of manufacturer, the War Office had never laid down exact requirements for the manufacture of its optical devices. Instead, a broad specification was given to contractors, with only key aspects being precisely laid down. The result had been that apparently similar items made by a number of contractors were unlikely to share any interchangeable parts, the manufacturers being free to follow their own route to produce an item that delivered the broad requirements demanded. In the case of the most widely issued pattern of binocular, although every maker's product gave exactly the same magnification and field of view, all of them differed in the make-up of their optical systems. In consequence, the difficulties of inducting new manufacturers were often compounded.

Even the largest and most competent of firms found problems in dealing with this lack. In August 1916, after lengthy and detailed negotiations, Barr & Stroud were given a contract to make 200,000 binocular prisms for supply to other firms who were already making complete instruments.[78] This was a novel move intended to remove one of the greatest production bottlenecks facing the OMGD, but the 'specification to govern manufacture' prepared by the War Office gave no information whatsoever about the form the prisms should take, beyond requiring that they be made of boro-silicate glass. The firm consulted the OMGD for the necessary dimensions, but the only information its technical branch could provide was to confirm the required glass type and the permitted tolerances on the prism angles. The actual dimensions of the prisms were not recorded by the War Office Contracts Department and the best suggestion by the OMGD was that 'It would be advisable to obtain from Ross two sample prisms to be forwarded to Barr & Stroud, from which they could take their own measurements'.[79] No prisms were apparently forthcoming from Ross, and eventually one of their complete binoculars was sent directly from Woolwich Arsenal. The OMGD made no attempt to record prisms' measurements and it was Barr & Stroud who finally provided the dimensions for incorporation in the contract to manufacture.

Having overcome what ought never to have been a problem, the company then produced the necessary jigs and set up the tooling to produce prisms on a scale far larger than ever done previously in Britain.[80] Then, doubtless to its consternation, the firm found that its two clients (Kershaw and the Brimfield Optical Co.) actually made binocular bodies that required prisms of slightly different heights, neither of which were the same as those provided by the sample binocular. The need to finish the prisms to different dimensions complicated manufacture and greatly slowed production until eventually Barr & Stroud took matters into its own hands and collaborated with the larger client, Kershaw, to produce a compromise prism that would actually fit both firms' bodies. This was done through relaxing the tolerances that had originally been provided by Barr & Stroud for inclusion in the contract. Deliveries went directly from Glasgow to the two binocular factories where the prisms were fitted and the finished instruments were then sent, as usual, to Woolwich Arsenal for inspection. But in October 1918 the Ministry's own Inspector of Optical Supplies became responsible for prism acceptances, and batches sampled began to be rejected because they failed to conform to the dimensions agreed with Barr & Stroud in 1916. Only after a series of letters occupying a month, in which time no prisms at all were sent to either firm, causing binocular manufacture to cease altogether, did the Inspector finally agree to accept the size that had been working perfectly for over a year, allowing the assembly of binoculars to be resumed.[81]

Another instance of problems from the failure to standardize designs appeared when Barr & Stroud was asked to take on the manufacture of an Admiralty gun sighting telescope in July 1918.[82] This had been made previously by Ross and by Watson and, unlike the War Office items, there actually were official drawings for it which were sent to Glasgow, along with a telescope from each firm. Difficulties immediately become apparent, because not only were the two samples distinctly different, but each had its own set of drawings and neither conformed to the Admiralty ones. A further complication was that the Watson sample telescope, which Barr & Stroud thought potentially quicker and cheaper to make, did not actually conform to Watson's own drawings. Furthermore, it used five different types of glass in its seven-component optical system, a degree of complexity which Barr & Stroud considered wanting of any good reason. The outcome was a complete redesign which, although producing a sight capable of being made quicker and at lower cost, contributed to the proliferation of different patterns that performed exactly the same job in the Royal Navy.

The failure to address uniformity in design and manufacture might seem to have been a singular omission on the part of OMGD, particularly when equally complex items such as the Army's service rifle achieved complete uniformity and interchangeability of components even when parts were made by numerous different contractors. But to be fair to both the department and the makers, equipment specifications were the province of the Services, and the official role of the Ministry was to facilitate production by contractors rather than to educate the War Office and the Admiralty in the art of planning for wartime procurement.

Maintaining Output

The OMGD's Weekly Reports show that its staff spent a large proportion of their time getting round problems for firms who were either too small or less able to take remedial steps themselves, as well as dealing with failures in larger, more experienced firms who should have been better able to manage their own affairs.[83] R. & J. Beck Ltd, whose principal Conrad Beck was closely associated with Frederick Cheshire's efforts to revitalize the optical industry, was criticized on 18 October 1917 for the poor overall quality of the company's output, and on 7 February 1918 for binocular deliveries that were so far in arrears that the contract 'might as well be cancelled'. At the same time, the firm was in dispute with the inspection department at Woolwich Arsenal, on which the OMGD's intriguing comment was 'It is impossible to condense all that our inspector has to say about this firm and the testing at Woolwich'. On Adam Hilger Ltd, for years before the war the country's leading specialist in prism manufacture, the comments were even worse; on 26 January 1917 the firm's binocular lens sets were 'unsatisfactory', on 15 February they were 'far from satisfactory', and on 1 November they were still 'not fit for service' and the entire lens workshop was so dirty that it 'must be swept out'. The Dublin firm of Sir Howard Grubb and Co. was far behind with its deliveries and on 28 February 1918 the OMGD noted that not a single telescope from a contract placed in October 1916 had yet been supplied, and for it to get back on schedule would depend 'on the help of supernatural agencies'. Broadhurst & Clarkson, a telescope maker established well before the war, had been proposed as a maker of Admiralty-pattern gun-sighting telescopes, but on 12 July 1917 the OMGD dismissed the suggestion as 'its plant is inadequate, its men unaccustomed to the work and ... the deliveries promised could not be made'. And in early February 1918 the Ross branch works at Mill Hill was said to be 'complaining bitterly that binocular bodies supplied by W. Watson & Sons for fitting with graticules were so dirty that it was impossible to work on them.

These reports are simultaneously telling and misleading. The OMGD staff making them were handling the day-to-day problems of firms who were, for the most part, trying to cope with large orders and the pressure to deliver quickly, whilst coping with shortages of hands and building extensions to their works. In August 1917 the Weekly Reports record ten firms with construction work going on, including Grubb, Hilger and Ross, most of whom were being criticized for tardy deliveries or poor quality.[84] The picture is one of struggle and, if not of failure, then at best of only limited success that prompts questions about how effectively the OMGD actually managed the wartime industry. But it has to be recognized that problems were bound to feature more prominently than successes in these records, and that lack of evidence of success is by no means evidence of its absence. Barr & Stroud, for example, features hardly at all except in relation to its problems with prism acceptances mentioned earlier in this chapter. There is little said about the massive output of rangefinders in Glasgow, although it could be asked whether the OMGD might have paid closer attention had the firm been in London, and provided more information as a result. It would be as unsafe to assume failure in the management of the industry from the records of Esslemont's administrative department as it would be to assume success for Cheshire's efforts on the technical side from his own contemporary claims.

Whatever the image created by the Weekly Reports, or the post-war printed history, there was no breakdown in the supply of optical munitions throughout the war, even before the Ministry of Munitions was created. Whether or not the OMGD chose the best way to organize the industry is debatable, and a case can be made that Cheshire in particular gave priority to the long-term interests of the country's optical instruments industry over the short-term needs to maximize output of munitions products. If he did so, his intention was to produce a viable industry that would be able to fill a role analogous to the private pre-war arms makers who had traditionally been expected by the state to make up the shortfall from the national arsenals in time of war, a goal that would in principle have been acceptable to his political masters in the Ministry. It cannot be said that he succeeded in that, largely because the conditions that provided his opportunity also combined to frustrate one of his two strategies; it proved much easier to set up a university programme in optics than it did to let the trade profit substantially in a material sense from its war work. His own departure from the Ministry meant that his ability to develop any provincial training scheme was largely eliminated, and the centre for excellence in optics remained a metropolitan phenomenon despite the national distribution of key optical munitions producers. As the next chapter shows, most of his efforts had little long-term effect on the optical munitions industry which, after all, always had remarkably little in common with the instruments industry.

Much of wartime optical industry can perhaps best be described as hermaphroditic. Those instrument making firms who were conscripted into becoming munitions-makers retained the characteristics of the former while acquiring those of the latter with varying degrees of completeness, which subsequently affected their success in their new role. That Esslemont's department was often unable to make sword-smiths from tinsmiths should not have been a surprise; that he was able to get any of them even to make table knives was an achievement in itself.

Although the contract records suggest that the bulk of useful output actually came from the small number of firms who were already experienced in optical munitions work before the war began, some of the newcomers did acquit themselves very well. The next stage in the industry's unfolding story looks at how well it performed up to the end of the war, taking as case studies three individual optical technologies and analysing how the conscripts fared compared to the most experienced member of the pre-war optical munitions community.

6 THE INDUSTRY'S WARTIME, 1915–18

War is the ultimate test of munitions supply. If a country's armed forces have inadequate quantities of munitions, or if they are significantly inferior to the enemy's, then the possibility of defeat increasingly becomes a threat. Even the greatest of command skills, determination and courage can be negated by the lack of the necessary equipment for the effective prosecution of war. In evaluating the success of munitions supply it might reasonably be supposed that the principal, perhaps the only, criterion of success is whether sufficient quantities of what was needed to ensure victory were indeed provided. But success in tests comes at different levels and assessing the degree attained by the optical munitions industry after mid-1915 is not one which can be measured by the simple yardstick of whether the armed forces received an adequate number of whatever instruments they needed. Equally important is to understand how it went about meeting the tasks imposed on it by the exigencies of war and, in particular, its relationship with the state in the form of the Ministry of Munitions' Optical Munitions and Glassware Department (OMGD). Because that body was inextricably linked to the industry from mid-1915, both must be considered together.

A convenient route towards understanding the evolution of the industry between then and the end of the war is to look at the processes by which the OMGD went about securing adequate output from the industry in what have been called 'individual optical technologies',[1] and examining how each progressed in the context of wartime conditions. Three of those individual technologies, prismatic binoculars, telescopic rifle sights and the man-portable single observer rangefinder, not only represent distinctly different sorts of instrument but also show disparate approaches to wartime production and demonstrate varying degrees of success.

Despite the importance of developing new products in response to the changing nature of warfare, volume production was usually the chief concern throughout the war and it was in achieving that where the greatest problems were encountered. The need to produce instruments in numbers never previously envisaged led to circumstances where the problems of obtaining sufficient factory capacity were regularly complicated by the emergence of situations where parts of an expanding production system either fell behind or became out of phase with others.[2] The presence of such difficulties, however, did not necessarily bring output to a halt even if its level was sometimes threatened and often reduced by them. Interim solutions were achieved through employing lateral measures to bypass problems in order to permit interim progress on a broader front. Such circumstances were found in all the three products just mentioned, but particularly in the ambitious moves to expand prism binocular production. Unlike the other two optical technologies, this effort was not entirely successful, largely because of a failure to synchronize fully all the elements of the production system, demonstrating the dangers inherent in some of the steps that had to be taken to create a substantially new industry at an accelerated rate.

Prism Binoculars

The prismatic binocular was needed in far larger quantities than any other item of optical munitions and posed by far the greatest problem of supply.[3]

The OMGD inherited a situation in which deliveries were already lagging far behind what was needed. The War Office ordered over 58,000 binoculars between August 1914 and June 1915, both at home and from French suppliers. That was more than all other optical stores put together and almost twenty times the number ordered in the financial year 1913–14, but was still inadequate for the Army's needs.[4] By mid-1915 estimated requirements had reached almost three times what had already been ordered, and output was so small that the shortage could only continue growing. The situation was considered to be so serious that the OMGD even organized the purchase of virtually every potentially suitable instrument from wholesalers and retailers throughout Great Britain, as well as asking members of the public to surrender privately owned ones.[5] That effort produced some 33,000 assorted instruments which, even assuming that all were suitable for military service, still came nowhere near addressing the longer-term problem. This continuing shortage presented a serious problem whose solution was to prove both protracted and elusive. The main difficulty, as in many other areas of munitions supply, was the lack of capacity amongst British manufacturers.

Although binoculars were regularly made in Britain before the war, their manufacture had been on a leisurely and relatively small scale that was far less than presently required. Ross, by far the largest British maker, had taken over fourteen years to produce no more than 25,000, averaging approximately 35 per week.[6] That binoculars could be made on a very large scale had already been amply demonstrated in Germany. C. P. Goerz in Berlin had reputedly made approximately 300,000 between 1892 and the outbreak of war in 1914, although it is possible that around 100,000 were a non-prismatic type produced for the Prussian Army forces.[7] Even if so, the firm had still made almost 180,000 prismatic binoculars in the time Ross had taken to make 25,000. Carl Zeiss at Jena had produced far more than Goerz, almost 433,000 between 1894 and 1914, averaging 50,000 yearly after 1910.[8] However, this was unique, and no other maker anywhere had made anything like that quantity or achieved such high rates of output.[9] The firm was extensively vertically integrated so far as the production of components went, even the optical glass for its lenses and prisms coming from its 'sister company', the adjacent Schott glassworks.[10] According to the British optical trade it had benefited substantially from the German Army's willingness to purchase and stockpile large numbers of prism binoculars, something which was seen as a kind of state subsidy.[11] With over 5,000 workers in early 1914, it was by far the largest optical manufactory in the world and regarded universally in the optical industry as a *ne plus ultra*. Nevertheless, the way the German firms made binoculars was no secret; the Zeiss works were regularly visited by representatives of foreign instrument makers and its production methods described in contemporary scientific and technical journals. Those methods, particularly for assembly, could in fact be replicated on a smaller scale in factories properly set up to perform the work, as Zeiss itself had already demonstrated with branch works in Austria–Hungary, Russia and England.[12]

A later claim, made after the war, that 'We knew [in 1915] how to make binoculars, but not … on a great manufacturing scale', was not strictly true.[13] Zeiss produced 1,000 binoculars weekly not by using secret or unique methods, but by employing large numbers of skilled workers to assemble an assured supply of suitable components, all of which were produced by the company itself in a large, modern and well-equipped factory. The real obstacle in Britain in 1915 was the lack of capacity for optical components as well as factory space and workers for their assembly. Although Aitchison, Ross, Watson and the London subsidiary of Carl Zeiss, Jena had all manufactured binoculars before 1914, none of them had the capacity to handle the unexpected volume now needed.

Whatever the scale of manufacture, the prismatic binocular was not an easy thing to produce en masse, even when the methods were familiar. Indeed, if mass production is defined as the use of fully standardized parts which are assembled into a finished object with a complete absence of hand fitting, then not even Zeiss accomplished that goal. The typical instrument used by armies and navies in 1915 used four prisms and at least ten lenses in two bodies joined together by a hinge mechanism. It required great consistency in its optics and, above all, care in final assembly if it were to function correctly.[14] Its lenses were machine polished in small batches, an operation still depending on a technician's judgement, but the assembly of the optics into their mounting cells and installation in the body housings was a hand operation. The final critical adjustment of aligning or 'collimating' the two telescopes in order to prevent the user from seeing a double image still had to be done individually, which meant using large numbers of workers to avoid a bottleneck and obtain high output rates. In Britain this was typically seen as skilled work to be done only by experienced craftsmen, but they were becoming increasingly hard to find by mid-1915. Many had already joined the Services and, as Cheshire had emphasized, there were no institutions to provide trained optical workers in substantial numbers. The existing makers accordingly faced real obstacles in trying to increase their output when they were asked to do so.

Ross and Watson were already heavily committed to both the Army and the Navy, making an assortment of gun sighting devices.[15] Despite having what was described as a 'beautiful factory' Ross operated on a much smaller scale than Zeiss or Barr & Stroud, employing only 320 workers in 1914 with a binocular capacity limited to around 100 per week in a mixed range of other optical munitions products which included the artillery dial sight and naval sighting telescopes.[16] Matters were little different with Watson which made gun sights and observation telescopes as well as medical and radiological products, and its binocular capability was much the same as Ross'.[17] The third binocular contractor, Aitchison & Co. had even less capacity. Its instruments were actually made by the Wray Optical Co. Ltd of Bromley, Kent, in which Aitchison held a minority shareholding.[18]

Wray, whose original business was making photographic lenses, had recently relocated to a newly built factory which had been set up largely in order to handle Aitchison's new War Office contracts, but its average weekly output was still only thirty instruments.[19] The new factory was five miles from the old one and inconveniently located for access by Wray's experienced workforce, some of whom had been reluctant to continue working for the firm when it moved and whose loss was hard to replace. Built in a small wood remote from any mains electricity supply, it depended for all its power on 'a single twelve horsepower gas engine' whose reliability was reported as uncertain, causing the entire works to come to a halt when it failed. Adding to labour shortages and interruptions, the factory was by no means self-sufficient for binocular making. There was no foundry for casting binocular bodies which consequently had to be bought-in from specialists who were already coping with orders from other makers, so that shortages compounded the problem of the factory's already limited capacity. Despite its recent opening the whole operation was too small to expand its output substantially without erecting larger buildings and installing extra plant and tooling, which would in turn require more operatives who would need to be trained, if indeed they could actually be found.[20] Even though the male labour force had been depleted by the massive scale of Army recruitment, Wray, like the rest of the largely London-based industry still endorsed the philosophy that only men could be employed for fine optical or mechanical work.

Carl Zeiss (London) Ltd, was the last of the binocular quartet. It was the British subsidiary of the German company and had been in a kind of commercial limbo since the outbreak of war.[21] Most, if not all, of its German supervisory workers had remained at the factory and the War Office had continued placing modest orders, but in early 1915 all the German nationals were interned and the business placed in the hands of a controller appointed by the Board of Trade, who then engaged Ross to oversee the running of the works. The extent of the factory's ability actually to manufacture binoculars, as opposed to assembling parts supplied from Germany, remains unclear but, for whatever reason, Ross never organized series production at the works. The components that were in the works when Ross took over were used up and the plant then largely used for the firm's other work. In 1917 some of its machinery was removed to aid the setting up of a completely new binocular factory in the north of England and the working practices associated with them were also taken up. Whatever the soundness of Ross's decision not to expand manufacture there, both the tools and the accrued expertise and their use were ultimately to be valuable in the efforts to expand binocular output, as will be described later in this chapter.

The main problem facing the OMGD was the lack of capacity in an industry that was quantitatively rather than qualitatively inadequate. That difficulty was not unique to British optical munitions manufacture. Even Zeiss, the world's largest optical producer, was unable to keep up with German demands. Although civil production was abandoned and the total workforce increased from 5,200 in 1913 to a wartime peak of 9,800, Zeiss could not maintain the substantive monopoly of government supply it had before the war.[22] Nor could Goerz, the next largest optical munitions producer, make up the shortfall. The German government was forced to draw other companies into its supply network for binoculars once it became apparent that not only was the peacetime capacity of its usual sources inadequate, but also the expansion programmes emplaced in 1914 and 1915.[23] From a situation of supposed surplus in August 1915 when, according to the Ministry of Munitions' official history the German government offered to exchange binoculars and telescopic sights for supplies of rubber, the German optical munitions industry progressively worked harder to keep up with its orders.[24] Despite having the world's largest optical industry, Germany still experienced problems with binocular output as the size of the Army increased and rates of attrition grew.[25]

At the OMGD, Esslemont had to devise a suitable strategy for the British problem. It became clear that a radical approach was needed because substantial short-term improvements in deliveries were unlikely. Efforts to purchase French binoculars were hindered because that industry was no better prepared than Britain's to handle huge orders and its own government was expecting to rely on it.[26] A promising source of supply in the US was thwarted because the War Office objected strongly to the design of the instruments being supplied by the Bausch & Lomb company. Here the War Office found itself between the proverbial rock and hard place. The objection came because the binocular did not conform to the constructional details specified for domestic contractors since 1909, something which at first sight might seem no more than quibbling and pointless bureaucracy in a time of acute need. However, in this instance, it was not necessarily the case.

All prism binoculars need to have some way of ensuring that the light rays emerging via the prisms through the eyepieces are parallel, otherwise double vision will adversely affect the observer. The British specification required this to be done by eccentric rotation of the object lens assemblies but Bausch and Lomb instead used very small screws to move the prisms relative to the light path. The objection was first of all that unlike binoculars normally made to the War Office's requirements the prisms were not fully supported in their seats and were liable to displacement through shock. If that happened, the instrument would 'squint' giving its user double vision. Additionally, not only was the efficiency of the method dependent on those screws remaining set, but the pressure points they generated led to stresses in the prism which were likely to lead to them chipping or breaking if subjected to shock or vibration. In a philosophy that saw durability in service as paramount, such a weakness was considered to be a serious obstacle. Despite achieving a delivery rate averaging 400 a week – then more than the capacity of all the British contractors together – on a contract for 20,000, the Chief Inspector of Optical Stores not only refused to move from the established technological paradigm but would countenance no further orders. Although the judgement was based on sound principles it can also be seen as misplaced when supplies were woefully inadequate. Certainly, it materially added to the supply problems of the OMGD, which lacked the power to over-rule him.[27] By the end of 1915 the problem was pressing hard and the industry itself was clearly incapable of providing a solution.

The OMGD was in favour of changing how the optical industry worked, not just in the context of optical munitions supply but for the future manufacture of instruments generally. Frederick Cheshire was convinced of the need for transformation through economic and educational changes directed at the existing firms which, in his judgement, were held back by a combination of educational and technological inadequacies. Esslemont, however, espoused a different approach, based on the premise that the root of the problem lay less in technological backwardness than in organizational shortcomings that could be overcome far more quickly. His radical solution proposed for 'the supply of binoculars under the scheme for development of home supply of these instruments' envisaged creating an entirely new factory in which would be concentrated the vast majority of British binocular production. The plan's novelty is emphasized because the idea for it originated outside the OMGD and the optical industry.[28]

During late 1915 Esslemont began discussions with A. Kershaw & Sons Ltd in Leeds, West Yorkshire, a business that was not in the optical manufacturing industry at all. Before the war the firm had successfully been making film projectors for the cinema trade, mechanical components for cameras, and complete camera bodies which were supplied to other companies for sale under their own names. None of those had ever involved optical manufacturing.[29] When the war began, the firm began making gun clinometers for the War Office and had been obliged to phase out civil products to cope with the increasing volume of government work.[30] By spring 1915 the business had effectively become a munitions contractor under the Defence of the Realm Act, a status confirmed when the Munitions Act became effective in July. Although the nearest thing to an optical component found in the clinometers was a glass-tubed bubble level, the Ministry of Munitions' classification of products fortuitously brought them under the control of the OMGD, and led to the firm's owner, Abraham Kershaw, being introduced to Esslemont. Although Kershaw had clinometer orders of 500 per week, the instrument was of relatively low value and uncomplicated to make. It could be made by many engineering firms, so there was no question of monopoly in supply, nor guarantee of an indefinite demand for it. What Kershaw wanted for his business was a munitions product of substantial unit value which would be required in large numbers. His motivation for this went beyond straightforward entrepreneurship because, having been designated a munitions contractor under the Munitions Act, his firm was locked into such contracting and now unable to return to civil commercial work.[31] By late 1915 the binocular supply situation provided exactly the opportunity he wanted.

Kershaw's ideas interested the OMGD, not simply because of what they might promise but also because he already had a substantial record of success. He proposed that high-volume binocular production could be attained by applying the principles and methods he had used successfully to produce complex articles such as motion picture projectors which combined high precision components with others requiring no especially high standards of manufacture. To make the former he had used dedicated machine tooling and enforced rigid standardization of parts, relying on a relatively small number of skilled machine operators and the total elimination of time-consuming hand fitting. For the assembly of those standardized parts he had been able to employ unskilled and semi-skilled female labour which was readily and cheaply available locally. Thus, a relatively small and inexpensive labour force of 200 workers was able to produce complex and precise items at competitive and profitable prices.

Kershaw had a purpose-built factory less than four years old and fully equipped with modern machine tools, some of which were housed in an 'air conditioned and temperature controlled' environment to ensure cleanliness. He had financed this through selling a 47 per cent share in his company to the Marion Co. Ltd, a London photographic wholesaler for whom he already made camera bodies to Marion's own designs.[32] In exchange for transferring some trade names and a monopoly use of some of his patented designs, Kershaw obtained a substantial cash injection and a guaranteed buyer for those designs whilst retaining control of the business. That seems to have been much more to his benefit than to the Marion company's and marked him out as very much the man of 'push and go' that typified the then current ethos at the Ministry of Munitions.[33] Despite his enthusiasm for applied technology and his business acumen, Kershaw had no academic qualifications or training of the type being advocated by Cheshire as essential for the advancement of the optical industry. In the event, that lack seems to have been no impediment to his relationship with Esslemont who recognized and welcomed a degree of entrepreneurial energy often lacking elsewhere in the optical industry.

Kershaw's combination of character and ideas suited Esslemont's immediately pressing needs, as well as Cheshire's wider aims to improve the optical instrument trade. The key element in Kershaw's plan was the extensive association of women workers and automated methods which he claimed would allow an eventual output of 1,000 binoculars every week, equal to the output of the Zeiss works immediately before the war. He was confident that this could be done successfully even though hand-adjusting would be involved in the final assembly of the instruments, a task which he also proposed would be done by semi-skilled female labour. His proposal also accorded with the Ministry's desire to sidestep labour practices that restricted output and was firmly committed to the 'dilution' of labour. One of the main purposes of the Munitions Act had been to secure the agreement of trade unions to relax the restrictive practices that excluded unskilled and semi-skilled labour, particularly women, from craft trades, but the largely London-based optical industry was firmly against introducing female workers, arguing that their training would be difficult and attempts to use them seriously counterproductive.[34] The OMGD had constant difficulties overcoming this argument about maintaining output, particularly as Cheshire's pleadings to keep the trade outside control used the same underlying logic of maintaining production at all costs, and its efforts to dilute workforces in London had met little, if any, success. But Kershaw was not in London, not involved in the optical trade and already used a mixed labour force. His proposals came without any impedimenta and provided not only the possibility of easing a critical problem but also the welcome promise of creating a precedent to weaken the general opposition to the widespread introduction of women into the established optical companies.

Kershaw's approach could not have been better timed or structured. Esslemont was willing to move outside the prevailing optical industry paradigm which considered instrument making to be the province of experienced skilled workers, and was happy to introduce and exploit this new initiative. His thinking was more broadly based than Cheshire's, which remained concentrated on transforming the existing trade for its future benefit rather than creating a new purpose-designed war industry. Esslemont asked Kershaw to prepare a schedule and details for the construction of an entirely new factory and its necessary tooling, and to define his ideas about how best to employ both the machinery and labour in it. Subsequent events illustrated what could be done when the Ministry of Munitions provided what was 'more or less a "carte blanche" budget'.[35] Unlike the Aitchison/Wray expansion in 1914, this venture benefited not only from a very large state subsidy, but also significantly better planning. Importantly, Kershaw sought and found a suitable new site less than a mile from the existing factory because he needed to retain its workforce to form the nucleus of the newer, much larger, one. That expansion would be facilitated because the works was located in a populous area well served by public transport and already providing a large pool of female labour for nearby large clothing factories. Kershaw hoped to be able to tempt those workers by the promise of more interesting jobs in a better working environment, an expectation which was seemingly well met once the factory was opened.

The site purchase was completed at the beginning of April 1916, and one month later a formal agreement was signed under which the Ministry provided what was described as a 'grant' of £20,000 for the project. This was by no means an outright gift and carried with it some stern terms and conditions.[36] Half the money was to be repaid, and as security a charge was taken on the land as well as the buildings and plant to be erected on it.[37] The conditions included maintaining a 'technical and commercial staff sufficient to ensure the manufacture of binoculars in the most scientific and skilful manner possible' and 'to train and use the service of unskilled and female labour to the utmost extent possible', both of which were implicit in Kershaw's original proposals. For Kershaw, this was a deal that was potentially even better than the one with Marion in 1910, and which emphasized the symbiotic nature of his relationship with the OMGD. The firm's share of the starting costs could be offset against high wartime taxation, the expected volume of business was considerable, and when the war ended Kershaw would have a large factory completely equipped for large-volume, high precision optical and mechanical engineering. In return, the OMGD hoped to obtain a resolution of its binocular problems.

The contract issued to Kershaw in mid-June was the largest single binocular order placed by the OMGD during the war, but its wording suggests that it was meant to be the first of a series of similar ones. It called for 25,000 instruments to be delivered at an escalating weekly rate, reaching at least 600 by the end of October 1916 and subsequently continuing at a thousand.[38] The Ministry's records show 'running' contracts regularly followed initial orders, often prolonging the first one for several years.[39] Contract 94/T/1039 also points to a condition in the optical industry on whose solution Esslemont was forced to depend for the success of the Kershaw project, which was itself intended to form the major part of a broader scheme to increase binocular supply.

Esslemont's 'broad front' was the increase of binocular output, but his immediate 'reverse salient' was the inexperience of Kershaw, both with making optical components and assembling them into complex instruments. The production of 1,000 binoculars a week required the presence of 10,000 individual lens elements and 4,000 prisms, besides 2,000 individual body castings, 4,000 lens cell assemblies and similar numbers of other simpler metal parts such as hinges and screws.[40] Although much of the mechanical work was broadly familiar, teething troubles were expected with the lens grinding and polishing and the final adjustment of the binoculars, all of which would be new to Kershaw. The OMGD expected pilot work to begin in the existing factory, and to provide the firm initially with optical sets, Esslemont intended to use experienced lens makers who had enough spare capacity to make binocular optics. These would feed Kershaw until its own output first provided self-sufficiency and then continued growing further to supply lenses to another factory which the OMGD proposed as a second new producer of binoculars.

The OMGD identified six potential lens suppliers, and the department's technical advisors liaised with them in developing suitable optical sets.[41] Cooke's of York, The Guaranteed Lens Co., Adam Hilger, the Hummel Optical Co., J. & H. Taylor and Taylor, Taylor & Hobson were engaged to make standardized eyepiece and objective lenses after June 1916. All of them found it unexpectedly difficult to satisfy the rigorous inspection criteria imposed by the Army's Chief Inspector of Optical Stores at Woolwich (CIOS) and there were delays in obtaining enough even to supply Kershaw's initial requirements. In February 1917 Taylor, Taylor & Hobson were finally delivering enough sets to allow Kershaw to begin assembling complete instruments in the recently finished factory (see Figure 6.1).

Figure 6.1: Kershaw-manufactured 'Binocular No. 3', 6 power. Manufactured in 1917 soon after the factory began production. Author's collection.

The capacity to make and assemble bodies was, however, far in excess of the supply of optics. By April, just 491 binoculars had been accepted by the CIOS, and Kershaw still had to reach the stage of being able to produce acceptable lenses itself. Only one of the six lens makers had reached an acceptable standard, and the lack of lenses was threatening to cripple the binocular expansion programme.

Esslemont's 'reverse salient' had not been dealt with, for reasons largely to be found within a second problem whose solution had been presumed when the Kershaw project was set in motion. The principal cause of delays lay not just in the ability of the lens makers to make lenses to the correct curvatures, but also in the supply of optical glass, whose quantity and quality was still by no means assured. Despite the progress with Chance Brothers achieved by Esslemont's efforts and described in the preceding chapter, the quality of certain glass types was particularly difficult to ensure, and during 1917 the inconsistency of those needed for binocular lenses was a constant source of trouble.[42] These problems were mainly responsible for holding back output of binocular lens sets in the first part of 1917, and were only eased by the OMGD's eventual decision to reach a short-term solution by purchasing suitable alternative glasses from France, and then by the subsequent improvement in Chance's quality control.[43]

Matters began to improve after May 1917, when a Kershaw trial lens set submitted to the OMGD was judged 'remarkably good and quite up to the standards of any [yet] submitted'. A week later, the OMGD finally passed a similar trial set from Cooke's, noting that this extra source 'should ease the situation so far as Kershaw's troubles are concerned'.[44] From then on output began to grow, with increasing numbers of binoculars being assembled using Kershaw's own lenses besides sets from Cooke's and Taylor, Taylor & Hobson. By October the factory was getting into its stride using production methods copied from the former Zeiss works at Mill Hill, as well as numbers of specially built machine tools designed in Kershaw's own drawing office, based on samples obtained from Mill Hill where they had been redundant, probably since 1915 and certainly since Ross bought the works from the Board of Trade earlier in the year.[45] By the autumn of 1918, weekly output had reached 800 and was still increasing when the instructions to scale down production were issued by the Ministry when the war ended as part of the process of industrial demobilization.[46]

Although delayed in reaching its goal, the Kershaw binocular project was eventually a successful collaboration between the OMGD and a firm that was innovative and soundly organized. It proved that Kershaw's faith in the transferability of his working methods from cinema projectors and camera bodies to binoculars was entirely justified. The delays caused by glass problems were outside his control, and it is clear that once those deliveries improved, the factory was able to work up to the high production rates that had been promised at the start. Dilution was achieved on a level not found anywhere else in the optical munitions industry, particularly in women workers. Where Ross employed around 17 per cent female labour and Barr & Stroud 16 per cent, four-fifths of Kershaw's workers were women, a substantial proportion of which was 'girl labour' aged under 18, a figure which 'seemed to astound the optical trade at that time'.[47] The earlier claims by London firms that women would create catastrophic wastages in lens making were refuted at Leeds, where 'very young' girls were trained to operate polishing machines within a matter of hours.[48] The cost of a Kershaw binocular was lower than from other makers, and by October 1918 the OMGD was proposing to discontinue lens deliveries from Cooke's and Taylor, Taylor & Hobson, partly because of Kershaw's then greatly increased output and partly because costs at Leeds were expected to be lower than at the other contractors.[49]

Kershaw's success, however, was not duplicated elsewhere in wartime binocular manufacture. It was the only commercial company set up during the war especially to manufacture them, but five existing businesses became newly involved in their production, and one state-sponsored assembly factory was started from scratch.[50] Sherwood & Co. appeared in 1915 in response to the War Office's assorted demands for optical munitions and continued to produce about fifty binoculars a week until the armistice of 11 November 1918, when soon afterwards it ceased trading. The OMGD asked Beck to make binoculars in December 1915, but the contract was subsequently suspended to increase dial-sight output and, according to the firm, none were actually made. (Beck's recollection was clearly wrong, because the author owns one such instrument.) Dollond & Co. began production in mid-1916 producing about twenty a week, and the ophthalmic lens makers Theodore Hamblin Ltd took an order for 2,000 in June 1916 and seem to have done atypically well, having completed it by 1918, after which they began making them for the Admiralty. H. F. Purser Ltd began production in February 1916, making around twenty per week until the end of the war. E. R. Watts & Son had an order for 1,000 in January in 1916 and proceeded to make them at the leisurely average of only five each week, perhaps qualifying them for the 'wooden spoon' award in being the least efficient of all the binocular makers. The amounts these firms added to output were minuscule in the context of total requirements. Apart from Sherwood, all seem to have been induced to take up production at the time when the OMGD was faced with a very large shortfall in output and when any new supplier was a welcome prospect. None of them, except possibly for Hamblin, ever had any large potential, but the final newcomer, the state-sponsored factory J. Brimfield & Co., was originally intended to do much better.

Esslemont had intended that the lens makers feeding Kershaw would quickly divert deliveries to another new factory intended to supplement Kershaw's production. This involved a different approach, with the Ministry setting up a new company known as J. Brimfield & Co., not to manufacture parts but to be a central assembly station for binocular components. The costs of equipping and fitting out the London factory were met by the state, and all its raw materials provided free of charge. Brimfield was to invoice the Ministry for assembly on the basis of labour and overhead costs plus a fixed percentage for profit. Although smaller than the Kershaw plant, its output was still intended to be considerable, and 5,000 binoculars were ordered at the end of July 1916, to be delivered at an increasing rate, with a minimum of 100 weekly being attained in less than two months.[51] The surviving records show that even the minimum rate was never attained. The plant was dogged by the same lens delivery problems that had affected Kershaw which, combined with poor mechanical quality, led to very high rejection rates by the CIOS.[52] Output seems to have been very small, and by the end of the war Brimfield was only taking in enough prisms to make twenty-five binoculars a week, a rate no better than many of the private contractors.[53]

Brimfield's failure to live up to expectations was almost an irrelevance by the armistice. The firm was never intended as a permanent part of what Cheshire hoped would be an improved post-war instruments industry, and Kershaw's increasing efficiency tended to make the venture redundant. Binocular production in the war had been something of the proverbial curate's egg both for Britain and her allies, and it is certain that far fewer than the 300,000 binoculars 'demanded' from domestic and foreign suppliers were actually delivered by the armistice. However, by then a factory to make them in Britain on a scale previously unimagined was running with increasing efficiency and the worst that could be said of binocular production was that high volume production came better late than never.

The Telescopic Rifle Sight

Volume was not the problem with the telescopic rifle sight. This was an instrument not found in the British Army before 1915 and its production on the scale of the prism binocular was never required. Intended for use by specially selected soldiers on a small scale, it presented a very different procurement problem to the binocular.[54] Previous accounts of its inception and manufacture have been misleading, in particular the reiteration of the assertion in the Ministry of Munitions' history that the industry was unsuccessful in producing telescopic rifle sights until the OMGD and the National Physical Laboratory jointly attacked the problem in 1917.[55] The device was actually introduced in early 1915 with regular small-scale production actually pre-dating the creation of the Ministry of Munitions, with output subsequently geared to escalating demand well before 1917.[56] The telescopic rifle sight, or riflescope, was one of the most successful applications of optical munitions technology during the war, and showed that the wartime industry was capable of meeting design and production requirements when an appropriate manufacturing infrastructure existed. Unlike binoculars, where the production capacity had to be newly created, the rifle-scope was needed in small enough quantities to be manufactured within the available capacity of the wartime optical community.

In late 1914 the War Office issued a general invitation to tender for a telescopic sight to be used with the Army's service rifle. This was in response to the German employment of specially selected marksmen using such sights which was causing mounting casualties and significantly affecting morale in the absence of any satisfactory means of retaliation. The requirement came at a difficult time for most of the optical industry because no such instrument was being made in Britain on a regular and organized basis, and growing demands from the War Office were increasingly occupying its capacity. Imported riflescopes had certainly been sold through the gun trade, some with British makers' names on them, but there was no domestic source immediately able to supply any type in quantity.[57] Although riflescopes were similar in concept to artillery sighting telescopes, those were specifically designed to fit mounting cradles on field guns and so large and heavily built that they were neither suitable for nor adaptable to the new requirement. Unlike either binoculars or rangefinders, the immediate problem was one of design.

Although all the optical munitions makers were capable of devising a riflescope, only two companies actually submitted designs judged worth adopting by the War Office, and neither of them had any previous connection with optical munitions. Aldis Brothers of Birmingham made photographic lenses and the Periscopic Prism Co. of Camden Town, London, chiefly produced lenses and prisms as components for the optical instruments trade. Neither had any government orders in late 1914 and unlike most of the industry both were actively seeking optical work. Coincidentally, each had already done subcontracting for Barr & Stroud, and both had been rejected by that firm because of the inadequate quality of their deliveries.[58]

Despite its adverse testimonial from Barr & Stroud, Aldis was nevertheless well regarded for its camera lenses and appears to have been far distant from Cheshire's depiction of an industry whose workshops were allegedly little more than shanties and whose principals were singularly lacking in higher education. The business had begun manufacturing in Sparkhill, Birmingham in 1902 and was owned by two brothers, both of whom who were mathematicians and graduates of Trinity College, Cambridge. The elder, Lancelot, had previously worked at the Dallmeyer optical works and was already an accomplished lens designer familiar with the latest in glass technology.[59] His younger brother, Arthur, joined him in partnership soon after being elected to a fellowship at Trinity in 1901, then also worked briefly for Dallmeyer before returning to Birmingham where he became interested in automated methods of lens manufacture. He subsequently spent time in Germany studying production methods, which resulted in the partners investing in automatic lens polishing machinery bought from the Ahlberndt Co. of Berlin. By 1912 the business was doing well enough to start building a new factory equipped for both optical and mechanical work at Hall Green, Birmingham, away from the polluted city atmosphere that was interfering with the delicate work of lens grinding and polishing.[60]

The new works was completed just before the war began, and by 1915 Aldis Brothers' catalogue included twenty-seven different photographic lenses and two for photo-micrography, all designed and made wholly in-house.[61] The OMGD was subsequently 'agreeably surprised' to learn that the firm had both a 'large and well equipped lens factory' and a 'scientific staff' able to tackle the problems of optical design.[62] The brothers' design of the riflescope can certainly be seen as endorsing Cheshire's emphasis on the benefits of scientific training in optics and the combination of a thorough grounding in mechanical engineering with a modern well-equipped factory. Within weeks a design, which was certainly not a copy of any existing type, was finished and satisfactory samples provided for the War Office, resulting in an order for a first batch of the 'Pattern No. 1' riflescope in January 1915.[63] Two hundred had been supplied and accepted before July when the OMGD took over responsibility for deliveries.[64] By then, Aldis had already suspended photographic lens production and, like Kershaw, become a de facto optical munitions contractor. In early July the OMGD contracted to take the 'entire output' of the works at a minimum rate of sixty weekly, an arrangement that was increased when Aldis later expanded and began making other types of sighting telescopes.[65]

The haste with which the sight was designed, and its designers' unfamiliarity with weapons may have been responsible for a number of mechanical shortcomings which affected the instrument's durability and caused problems in service. In particular, the vibration of firing loosened the range-adjusting mechanism, and penetrating moisture caused fogging. Either rendered the sight unusable and when the problem was referred to the OMGD's Technical Section it was recognized that substantial redesigning was necessary to eliminate them. The Aldis brothers may have been highly competent in computing lens systems, but they had no previous experience of the problems likely to be encountered by a telescope rigidly mounted to a powerful rifle, subjected to considerable shock every time the weapon was fired and regularly exposed to inclement weather. The Pattern No. 1 had been put into manufacture without any proper engineering drawings, doubtless because of the haste involved and because the War Office simply did not require them. As usual with optical munitions, the makers were left to decide for themselves the details of design and the method of manufacture; if they chose to do without detailed drawings then that was solely their concern. By late summer 1915 when the future requirements were recognized as likely to be substantial and long term, the OMGD decided to standardize on an improved version better suited for larger-scale production. Its Technical Section liaised with Aldis about the necessary changes resulting in satisfactory prototypes of the revised design being accepted in early November and finalized production drawings were issued by the Technical Section less than three weeks later. Manufacture then proceeded steadily without further significant technical problems until a decision by the War Office in 1918 to adopt a completely new design based on a captured German one. The riflescope may have been perceived as inferior to the one chosen for its replacement but nevertheless it was adequate for the job and illustrates that 'best' is not necessarily a critically important role in optical munitions.

Aldis was an excellent example of how well the wartime optical munitions industry could perform. By building on a base of existing sound scientific and technological practice, and drawing on the resources of the OMGD to bridge gaps in its own pre-war organization, the firm successfully moved from being a small-scale civilian maker of photographic lenses to become a large and highly specialized maker of complex service optics. Its factory was extended twice during the war and by the armistice it was six times the area of 1914, producing not only riflescopes but derivatives of them for aerial gunnery, as well as large numbers of artillery sighting telescopes. Experience gained with them was used to design a novel electrical Daylight Signalling Lantern (the 'Aldis Lamp') which used a reflector and lens system to create an intensely bright point source of light that could be seen at long range even in the strongest daylight. The lamp needed to be aimed precisely to permit the observer to see it, and an inexpensive but durable aiming telescope was designed by the company especially for it. By late 1917, the firm's capacity was large enough to allow the Aldis brothers to compute new lenses especially for high altitude aerial photography and to begin their production on a substantial scale. The spin-off of munitions contracting provided Aldis with the basis for many of its post-war civil products; the range of photographic lenses was extended and the Aldis Lamp sold widely to both merchant shipping companies and foreign governments.

The second riflescope maker, the Periscopic Prism Co., is less well documented. Like Aldis it was quick to produce a design, partly because it also lacked orders, and partly because one of its directors, A. B. Rolfe-Martin, combined optical design skills with an interest in rifle shooting. Rolfe-Martin's shooting experience might have been expected to give his firm an advantage, but the OMGD decided to keep both the Aldis and Rolfe-Martin design in production, suggesting that neither showed any distinct superiority. Even if the OMGD thought the manufacturing quality of its riflescopes was less good than the Aldis sight, it was numerically the larger maker and its products were still adequate for the tasks set. Over 4,400 had been made by April 1917 and production continued until the war's end, by which time the firm had been taken over by the state and was also making sighting telescopes for artillery.[66]

Riflescope production hardly taxed the industry at all. Those used through-out the war were made by just two firms who met the Army's requirements with relatively little trouble and marked a success for the optical munitions indus-try. Their importance to the psyche of front-line units was out of all proportion to the relatively small numbers made and the value of the contracts. The total of around £70,000 spent on the 10,000 or so purchased during the entire war was less than three-quarters of the first binocular contract placed with Kershaw, and only a twentieth of what the War Office spent on rangefinders with Barr & Stroud, but the instrument was recognized as vital to effective sniping and simply had no substitute. The War Office's eventual decision to replace both types with one copied from a German riflescope stemmed not from any crucial functional failure in what was available but more from a supposition that the proposed alternative was fundamentally superior in design and so capable of bet-ter standards of performance.

In tactical theory, the riflescope would have been employed most effectively by a team of three men using not only the rifle-mounted sight, but also a power-ful spotting telescope to locate targets and a rangefinder to determine distances precisely and set the riflescope accordingly. In service, though, the realities of trench warfare meant that even if targets still needed location the ranges were relatively short so that precise sight settings were not critical, and the need for the portable rangefinder as part of the team never materialized.[67]

The Rangefinder

The man-portable rangefinder was the most complex item of optical munitions to be produced in great volume during the war; yet, in an apparent paradox, it was the instrument that posed the least problems in supply. And, ironically, it may well have been the least important optical device in the conditions of the Western Front where the need to determine distances rapidly became rare once the static nature of trench warfare emerged. Unlike the other optical devices used by the Services at the start of the war, the capacity to mass produce range-finders already existed with Barr & Stroud whose records reveal much about its attitude to wartime contracting and the difficulties encountered.

Barr & Stroud's pre-war capacity was already considerable in August 1914, and the suddenly increased War Office orders at that time were, to some extent, off-set by the enforced termination of foreign business which enabled output more or less to keep up with growing demand. As orders continued to grow, the company expanded both its premises and workforce and kept pace with demand whilst simultaneously carrying on research and development into new manufac-turing processes and instruments, a pattern of activity not found elsewhere in the British industry. Barr & Stroud has been more closely associated with its dealings with the Royal Navy during the war,[68] but for much of the period it actually had a greater involvement with Army business, first directly from the War Office and later from the Ministry of Munitions.

Land service business was indeed substantial, being 80 per cent more in value than the Admiralty's during the war.[69] Approximately 19,300 rangefinders were supplied to the War Office, almost 16,000 of which were infantry ones and the rest a rather larger model made especially for the artillery.[70] Prices varied but little during the war despite inflation, with their approximate average values being £58 and £80 respectively, representing a total business of around £1,360,000.[71] The production requirements for each service's orders were quite different and handled separately. The Army's smaller rangefinders were needed in vastly greater numbers than the large naval ones and posed manufacturing problems which were essentially similar to the binocular. The rangefinder was far more complex, but most of its components could, in theory at least, be made rapidly, as could those for the binocular. Barr & Stroud's serious problems in getting enough optical parts in the first year of the war have been discussed earlier in chapter 4, but as its own lens and prism production increased the main factor restricting output became the assembly of completed parts, where large numbers of workers were needed to achieve rapid deliveries of completed instruments. Like the binocular, most of the assembly could be done by semi-skilled workers with relatively little training, but finding enough hands was still a problem and manning shortages grew throughout 1915.

By the end of the year the firm's general manager, Harold Jackson, was complaining that it was difficult 'explaining to athletic young men that ... making tiny prisms is as valuable to the Country as sticking a German with a bayonet'. As in other industries, it was hard to keep men away from the colours, and the OMGD promised to send Barr & Stroud an open letter explaining to the workforce how vital the firm's output really was to the war.[72] But neither the letter nor the posters subsequently displayed in the factory could persuade the young and not-so-young to eschew the bayonet and by spring 1916 labour shortages for the more precise operations were sufficiently serious to require some urgent remedy. The only new source of labour available had never been used previously by Barr & Stroud, to whom the idea of having women in the factory was a novelty which was approached with some considerable forethought.

Unlike most of the London optical industry, it is clear that the firm had no prejudice against women, but where Kershaw employed them for 80 per cent of its workforce, Barr & Stroud never had more than 16 per cent on its payroll.[73] That smaller proportion resulted not from any reservations about their likely abilities, but from the specific nature of the work on which the company planned to employ them.

In contrast to Kershaw's scheme of operations, where 'girl labour' was widely employed and where female workers were expected to be able to carry out straightforward machine-minding tasks within only hours of starting work, Barr & Stroud intended to utilize women on quite different tasks in the more complex process of producing rangefinders. Those included particularly fine and skilled operations such as scale dividing and engraving whose accuracy was crucial to the instruments' accuracy in use. In another major difference, the company planned not to poach female workers from their present employment as factory workers but specifically to recruit from women who had either never worked outside the home or had been employed in retail or similar trades.

Barr & Stroud predicted that its first, and possibly greatest, difficulty in employing women would be in their adjustment to a factory environment that was very different to anything they had known previously. Those engaged were selected by interview from what was described as 'a large number of applicants' and it was originally intended that they would 'go through a short course of training ... to accustom [them] to factory life' before instruction began in a special training school teaching skilled operations such as milling, scale dividing and even tool grinding. This was far removed from the earliest private initiatives to draw in female labour elsewhere in the munitions industries, when, for example, the aristocratic Ladies Moir and Cowan had begun organizing training for other similarly 'leisured ladies' for part-time weekend work at Vickers' Erith works in Kent.[74]

Barr & Stroud saw women as potentially an asset to be directed at specific skills shortages, but they were by no means the first Glasgow munitions makers to adopt such ideas and might possibly have been encouraged by the steps taken in 1915 by William Beardmore & Co. to train women for skilled engineering work.[75] Even though Barr & Stroud's ambitious induction procedure and training programme had to be curtailed in deference to the Ministry's policy of not delaying the 'introduction of large numbers', women were still given individual instruction in the optical and fitting shops where their performance was judged 'as efficient as men'. Unlike other Glasgow engineering firms, including the pioneering Beardmore which took a considerable time to provide even the most basic facilities for women,[76] Barr & Stroud planned their integration from the outset and went to considerable lengths to provide adequately for it in order to maximize the benefits of the new labour force. Separate 'cloak room' facilities were provided in 1916 for the 300-plus women workers expected, and a full-time 'Matron' appointed to look after 'the women's welfare and interests in every way'. According to Harold Jackson's post-war summary of the exercise, there were few difficulties in the workshops themselves, and the quality of work done by relatively inexperienced female operatives was generally highly satisfactory, with their attitude to the work 'more assiduous' than their male colleagues.

Labour problems at the factory were managed more easily than those connected with optical glass. Its shortage and inconsistent quality was a different sort of problem, and although never so bad as to bring production to a halt it caused difficulties right up to the war's end. In late 1916, quite independently of other optical glass initiatives by the Ministry of Munitions, Barr & Stroud began small-scale production of the types of glass that had proved the most problematic to obtain during the war. That was part of a general policy to make the firm increasingly independent of the outside supply of raw materials and components, measures which were so successful that by 1918 almost the only important remaining external requirement was for the most common glasses that were required in bulk. Deliveries of these were still difficult even in late 1918, but by careful management and complaining regularly and vociferously to the OMGD, stocks usually remained high enough to keep production going.

Rangefinder supply ran continuously throughout the war. Both the Army and the Navy were adequately supplied and Barr & Stroud also had enough capacity to overhaul large numbers of instruments damaged in service, running and staffing a workshop at Woolwich just to handle Army work. Far from being worn out in November 1918, the firm was further expanding its works to handle new Admiralty orders and was undertaking optical computation on an increasing scale. Where in 1914 the vast majority of optical components were bought in, four years later the business was entirely independent of other sources for grinding and polishing and so well equipped to handle large-scale production that it was ready to begin supplying infantry rangefinders to the American Expeditionary Force which was then being deployed in Europe.[77]

Despite the success of rangefinder production for both services, it was initially seriously inhibited by the War Office's attitude to the company. Soon after the war began, the War Office began to act as a difficult and obstructive buyer, interfering in Barr & Stroud's relations with its other customers. By mid-1915 matters had reached a point where a crisis in output was averted only by intervention of the Ministry of Munitions which acted as a buffer between the two. The subsequent relationship between the company and the OMGD provides an alternative to the picture painted in the official history where the entire optical munitions industry is portrayed as the beneficiary of state assistance.

When the War Office began its large orders in the autumn of 1914, it sought to impose conditions under the provisions of the Defence of the Realm Act that the firm considered so irksome and disruptive that they would greatly prejudice the prospects for output. Barr & Stroud's problem was that the War Office's stance was to demand priority for its orders over all other work on hand. The Defence Act did indeed give it the legal power to do so, but the intent of the legislation was to force contractors to prioritize war work over other commercial business; in Barr & Stroud's case all its other work was also government munitions contracting, most of it for the British Admiralty. The War Office seemingly remained oblivious to the problems created by such insistence, particularly when it sought to extend its primacy over Admiralty orders as well as those for both actual and potential allies in the war.

Demands from the War Office that its orders should be processed ahead of the Royal Navy's clashed with assurances that the firm had already given to the Admiralty. Even before war was declared, Barr & Stroud had assumed that its chief obligation would be to the Admiralty and had promised its full cooperation in providing whatever the Royal Navy needed. The insistence of the War Office on primacy was a problem, even if it could be worked around to a large extent, as the Admiralty's orders were unlike anything the War Office was calling for. Much harder to manage was the directive to curtail the export of rangefinders to France, Britain's principal ally. That instruction came despite long-existing orders and the French army having urgent need for them, to say nothing of the company's contractual obligations to deliver them. Then, in January 1915, the War Office infuriated Barr & Stroud by telling the firm to end its continuing, determined efforts to secure the United States Navy's future rangefinder business, in order to concentrate on output for the British Army. Objections were fruitless: 'The War Office has even threatened to take over the control of our works', wrote Jackson to the firm's American agents with ill-concealed bad humour.[78] The next month, in a display combining obduracy and ineptitude, the War Office totally forbade the shipping of any rangefinders to France, so disrupting an inter-governmental agreement to obtain essential supplies of French optical glass in exchange for uninterrupted deliveries of rangefinders. That brought swift repercussions from the French government which temporarily ended all supplies of the optical glass on which much of the other optical munitions output for the War Office largely depended.[79]

To the company, the War Office was an increasingly difficult and obdurate customer. Having ignored Barr & Stroud and its rangefinders until just before the war, it now demanded priority in all things, disregarded all the firm's commitments to the Admiralty and allies alike, frustrated opportunities for new foreign business, and yet still failed to take up everything that the firm had available for delivery. By June relations had deteriorated to the point where there was open distrust of the War Office's good faith, with Jackson seriously believing rumours that it intended to buy artillery rangefinders from Bausch & Lomb in the US rather than in Britain.[80] The War Office was far from an ideal client, and what must have been particularly irksome to Barr & Stroud was that the value of its business since August 1914 was still only two-thirds of the Admiralty's in the same period and far less than France's had been in 1913.[81]

The arrival of the OMGD ended this imbroglio and illustrates the pragmatic approach it frequently adopted. Cheshire and Esslemont both understood that, irrespective of their duty to organize supplies for the Army, the Royal Navy's need was equally great and could not be obstructed. The OMGD immediately instructed that dealings with the Admiralty should be done directly and independently of contracts for the War Office.[82] This permitted the firm to keep the Navy's business adequately prioritized and doubtless eased problems in proceeding with Admiralty work. Then, in July, Cheshire visited Glasgow, marking the start of what seems to have been an amicable and symbiotic, if once again somewhat asymmetrical, relationship between the firm and the OMGD.[83] As far as becoming an officially government-controlled establishment was concerned, the company showed no sign of opposition. Jackson had no doubt that de facto War Office control had existed since December, so that the prospect of dealing with a new and more accommodating agency was likely to have been an agreeable prospect.[84]

The correspondence between Cheshire and Esslemont and the firm shows that they regarded both the company and its senior staff as their organizational and intellectual equals, rather than seeing them, like the rest of the optical industry, as being seriously in need of assistance and direction. However, this sentiment was by no means wholly reciprocated, and the company's records indicate that it sometimes felt that the OMGD itself required instruction and correction. It may have helped the firm's confidence to know that almost its entire board of directors had better academic credentials than the OMGD's management and that the business was indisputably the world's largest and most successful specialist maker of optical munitions. Self-confidence was never lacking at Barr & Stroud although ironically its most forceful personality, Harold Jackson, had abandoned his studies some fifteen years earlier to work for Barr and was the only senior executive without an academic degree.

In August, Esslemont asked Barr & Stroud for 'private and confidential' details of its experiences with British optical glass, and Jackson's reply dealt at length with both the commercial and technical difficulties in dealing with Chance Brothers.[85] His detailed analysis of Chance's operation and strong assertion that unsatisfactory business structure rather than scientific inadequacy was the root of Chance's manufacturing problems went far beyond what had been asked for. It showed a grasp of systemic problems that was exactly what Esslemont needed to know, and it was a theme which he subsequently took up not only with the glass maker but also in his later approach to binocular manufacture. In September, Jackson took the initiative to call for a 'firms' conference' across the whole optical industry to examine the problems of overcoming the lack of optical glass and skilled optical labour, which Esslemont duly organized.[86] Jackson's summary of the current situation perhaps made uncomfortable reading for the OMGD. He first questioned Esslemont's continued confidence in guaranteeing adequate glass deliveries and reminded him that increased output at Glasgow depended on Barr & Stroud itself being able to develop 'machines and methods' for the successful use of unskilled labour, in view of which he firmly refused to commit the firm to further increases in production, whatever their importance or urgency.[87]

At much the same time he promised to help Frederick Cheshire to give a lecture on rangefinders by providing background technical information to fill gaps in his knowledge. Jackson sent the material and 'a little model' of a soldier using a Barr & Stroud rangefinder, with the firm's compliments and a gentle reminder that virtually every European army already had both the information and the model, including 'the Technische Militar Komittee [*sic*] in Berlin'. In a tactful postscript he added 'If Mr Esslemont is jealous [of the model], I shall be pleased to send him a duplicate'. Jackson perhaps considered that both of them were starting on what would be a steep learning curve.[88] Cheshire's lecture, largely constructed around the material provided by Barr & Stroud, subsequently won him the Optical Society's prestigious Traill-Taylor Memorial Medal for advances in optics, which doubtless helped cement good relations between him and the company.

It is difficult to see any other British firm taking these attitudes at the time. But then Barr & Stroud was in many ways the embodiment of what Cheshire was prescribing as necessary for the future of the whole optical industry – a soundly managed, well-financed and profitable business that used scientifically trained staff, modern plant and manufacturing methods, employed a motivated and well-managed workforce and which was, above all, demonstrably successful against foreign competition. It was therefore ironic that Archibald Barr himself did not see the firm as an optical instrument maker, or indeed any kind of scientific instrument maker; he was absolutely certain he headed a specialist engineering company that employed optical components in mechanical artefacts, and the war only served to confirm this long-standing conviction.

Barr emphasized this by refusing to connect the firm with other optical manufacturers. He repeatedly turned down invitations to have the firm represented at informal trade gatherings in London early in the war, and declined to join a proposed scientific instrument makers' federation in July 1915 because 'honestly (and privately) we do not feel disposed to place ourselves in any way under obligation to abide by decisions made by those who are possibly under very different conditions from our own'.[89]

This may have shown a measure of insularity or even amour propre, but it was still a realistic statement. Eventually in 1916 he drew a line under attempts to persuade him to join similar bodies by very firmly declining not only the offer of the vice-presidency of the newly formed British Optical Instrument Makers' Association but even membership of it. He was not 'disposed to associate' the firm because 'our work is so very different' from other optical makers.[90] With 1,500 metal workers to just 100 optical workers, both the firm and its work were indeed as different as the way it had achieved its success.[91]

The optical munitions industry's efforts in the three areas examined here embraced degrees of success that differed in their completeness and scale. The binocular and riflescope were both instruments not previously produced in great quantity or at high speed in Britain – indeed, the riflescope had at best only been made in minuscule amounts and possibly not at all. Both the binocular and the riflescope were manufactured during the war by firms who had no previous experience with them, the riflescope being made with much more conspicuous success largely because the numbers needed were relatively small and well suited to the sort of optical manufacture typical of British pre-war practice. But that practice was not the skilfully distorted image that Cheshire deliberately fostered to achieve a long-term goal, rather it embodied both scientific training and commercial expertise that could adapt well to a new specialized requirement when that need fitted in with the existing capacity. Aldis in particular built on its inherent skills to become a much larger entity, making a wider variety of instruments than it did before the war; it benefited from its war work and 'span' into its own growth. Success in volume binocular production was harder to find because the lack of any adaptable facility meant everything had to be created from scratch, and although clearly in sight by the end of the war the promised high output rate was still not fully achieved. The characteristic of the riflescope makers in being strangers to the product was emphasized in Kershaw's manufacturing background, and extended by a considerable degree through his complete unfamiliarity with optical work. But Kershaw showed that his production methodologies could be successfully translated into a different field and, had the supporting infrastructure that Esslemont gambled on been in place, the project would have doubtless fulfilled its promise sooner. Unfamiliarity was shown to be no bar to successful manufacture.

The rangefinder was a very different case to the other items. Barr & Stroud showed that the transition from peace to war production could still be problematic even when the business was both experienced and well equipped to handle large-scale output. This was partly because of the unexpected scale of the war, but partly because the firm's incomplete vertical integration left it vulnerable to a supply structure which was much weaker than the body it fed. The scale and organization of the business allowed a metamorphosis sufficient to eliminate most of its vulnerability and it reached the end of the war in a condition almost universally improved from mid-1914, able to design and manufacture a wider range of more complex instruments with a greater level of efficiency. Barr & Stroud needed little help from the OMGD, except where issues were essentially political such as War Office policy in 1915. If this chapter has said relatively little about problems in rangefinder manufacture for the Army, it is because there is really little to say; even if the job was problematical, the goods were delivered in sufficient numbers to satisfy the level of demand created by the client.

In early November 1918, the British optical munitions industry was working at high speed and full capacity, preparing either to increase output still further or to introduce new and improved instruments. The armistice signed on 11 November apparently took all the manufacturers by surprise and the reaction of the Ministry of Munitions amounted to an extreme case of emergency braking which threw the industry into such disarray that, if only for a short time, it completely lost control of its course. The next part of the story examines how its members coped with the traumas resulting from might truly be described as an unplanned and unanticipated event.

7 INDUSTRIAL DEMOBILIZATION AND IMPLOSION, 1919

For the optical munitions industry, the signing of the armistice on 11 November 1918 came unexpectedly and, in one sense, prematurely. The effect of the cessation of fighting on Sunday, 11 November had almost immediate repercussions for its business in the form of order cancellations on what amounted to a heroic scale. This had in no way been anticipated by the industry. Irrespective of the political desire to end hostilities and the sentiments being voiced in the press, it had been very much 'business as usual' right up to 10 November 1918.[1] There had been no scaling down of contracts and no warnings from the Ministry of Munitions of any imminent likelihood of cancellations. Indeed, firms were still being exhorted to step up production and new contracts were still being issued.

All the country's optical instrument manufacturers were wholly employed in the war effort and none had even begun to consider in practical terms any policies for industrial demobilization and a return to peacetime trading. Furthermore, the efforts of the Ministry's Optical Munitions and Glass Department (OMGD) to set the optical industry on a better footing had not yet reached a stage where substantial improvements in organization or infrastructure had been achieved. The pre-war optical instruments industry had been conscripted and undergone a metamorphosis into the wartime optical munitions industry with some considerable success, but nothing had been done to cater for the inevitable end of large-scale government orders. For the hugely expanded optical industry, the transition back to peace was a difficult process. Its constituent firms had to come to terms with the twin difficulties of the disappearance of government business and the problems of re-engaging with a civilian market from which they had been excluded since 1915. The 'demobilization' of the wartime industry highlighted the still incomplete state of the Ministry of Munitions' efforts to improve the condition of the optical industry, and demonstrated the collective inability of much of the industry to cope with the organizational and financial problems which the war had generated and its conclusion compounded.

The war had done much to create an expanded optical munitions industry which was recognized as essential to the successful conduct of the war. By late

1918, the industry had come to constitute an entire technological system of manufacture. The widespread changes imposed by the closing-down of wartime demand may have been an unavoidable outcome of returning peace, but they were by no means helpful in keeping together a nucleus of manufacturers. As firms made the transition back from wartime manufacturing, sometimes with much difficulty and little success, they had to cope again with a plethora of variables similar to those that had influenced their induction into the war effort. All had to attempt to manage a range of political, economic and social factors besides the scientific and technical ones relating directly to the artefacts they dealt in.

The Problems Embedded in Total Mobilization

The Great War had drawn into optical munitions manufacture firms which previously had little or no experience in the field. Their pre-war business had come from manufacturing an assortment of microscopes, survey apparatus, photographic lenses, telescopes and similar instruments, trade which had gradually vanished as the Ministry of Munitions involved more and more of those businesses in military manufacturing. The result was that these 'conscript suppliers' lost not only their pre-war business, but also close contact with the markets for that business. The Ministry's contract records show that the pre-war 'commercial' optical makers had become so involved with government contracting that by November 1918 they were completely dependent on military equipment for their business. Mobilization of the optical instruments industry was complete with every optical manufacturer operating in 1914 drawn into the war effort. Not only had those businesses put aside their familiar products, they had also been forced to adopt changes in their workforce and methods of production. The demand for instruments and components on a scale previously unknown, along with the necessity to use less-skilled labour, meant that new machinery had been introduced into the industry and what often amounted to an entirely new workforce had been created at the Ministry's instigation. In theory those rationalization measures should have placed the makers in a good position to move back into the commercial marketplace, but factors external to the industry stood to inhibit, or possibly nullify, the realization of some of those potential benefits. The successful substitution of semi-skilled labour to replace craft workers enforced by the 'dilution' requirements was likely to be compromised by government guarantees given to the trade unions in 1915 and 1916 that those who had worked in the industry were guaranteed re-employment at the war's end. Those returning from the war were ill-prepared for an often radically changed factory environment, even if there was work for them to do.

Although the optical industry had indeed been transformed by the war, this metamorphosis had been chiefly directed to meeting the war's demands, and in consequence little or nothing had been done to refine existing commercial products, let alone develop new ones. There were some exceptions where specific requirements had stimulated new optical products, such as photographic lenses for aerial reconnaissance, but these were very much the exception rather than the rule and, in any case, had no immediate application for the civil market. The potential for more efficient manufacture which had been embedded into at least some of the makers was, to a large extent, negated by four years of stagnation in design and civil product development, as well an enforced divorce from both domestic and overseas markets. Although there was indeed concern in the Ministry for the industry's long-term future, the measures to secure that future were still only partly formed by late 1918. The cessation of war-related orders left the conscript makers in a problematical situation where they had neither government business nor commercial trade to rely on.

Similar problems also faced the much smaller number of companies which had been most involved in optical munitions manufacture before the war. All but one of those 'regular' contractors had also made commercial products. Like the conscripts, the two largest, Ottway & Co. Ltd and the Ross Optical Co. Ltd, were suddenly faced with the problems of returning to peacetime trading in civilian markets. Barr & Stroud, however, had no civil products before the war and relied entirely on armament-related business whose greater volume and value had come from foreign states. The imminent ending of large-scale munitions contracts posed an even greater potential problem for Barr & Stroud than it did for the other wartime makers, all of whom had experience of the civil trade in optical goods which might be resurrected, even if with difficulty. For everyone, though, the immediate problem was disengagement from the dependency on government orders and then securing the best possible financial settlement for cancelled contracts.

The Ministry of Munitions and Industrial Demobilization

The placing of contracts during the war had been done under rules which safeguarded both the state and, to a somewhat lesser degree, the manufacturers. The regulations provided for amendments to contracts allowing their extension as well as their early termination. In the latter case, there was the right to compensation for the manufacturer, so long as it was not in breach of the contract's terms.[2] Many of the wartime contracts were very substantial and intended to run for lengthy periods before their completion. In some ways such employment was far better than the civil marketplace had ever offered. Optical munitions work for the state during the war was a steady source of business for which payment, if sometimes slow, was never in doubt, and for which firms were frequently offered financial inducements to modernize and extend their premises, machinery and workforce. Despite the apparent security and certainty of munitions work it was never thought of as continuing in perpetuity. The contractors should have understood that it was by no means immune to curtailment or cancellation and, most importantly, that the reason for it on such a large scale would not last indefinitely. The reactions from virtually all the industry after the November armistice shows the extent to which the pressures of war work had caused consideration of that ultimate eventuality to be postponed, disregarded or sometimes seemingly completely forgotten.

Contract cancellations began immediately after the armistice was announced, probably even on 11 November and certainly no later than the following day.[3] For the Ministry's Contracts Department, the war was indeed considered to be over. This rapid reaction to events was in line with the general policy that had been given 'much attention' by the Ministry of Munitions in the months preceding the suspension of hostilities.[4] The greatest problem identified by the Committee on Demobilization and Reconstruction which had been set up in connection with the war's eventual end was the cancellation of contracts no longer required 'with minimum disturbance to industry and labour', something which might well have been expected to be fraught with difficulties. The consequent necessary settling-up of contractors' accounts had been acknowledged as inseparable from that process and recognized as the next most important matter to be dealt with.

Two alternatives for the termination of orders had been considered at length before a decision was reached that was intended to be applied universally across all munitions production. The first possibility was to begin by slowing down the tempo of output and then to reduce the scale of production so that contracts would go through a gradual process of arrest, spreading the rate and scale of redundancies over enough time to let civil production resume and absorb the munitions workers who would be progressively released from war work. That process would be helped by what was expected to be high levels of demand for consumer goods after the shortages of the preceding three years. Had the optical industry been involved in the Committee's deliberations and so directed to think of its own post-war future, that suggestion might have been seen as a practical, if not necessarily comfortable, route for the transition back to peace. The second, and from the industry's viewpoint a more radical and difficult option was to discontinue munitions contracts 'at the earliest possible moment' whatever the immediate effect on the labour market. Such an adventurous, not to say risky, policy was prompted by the expectation that rapid freeing-up of capacity for meeting the anticipated pent-up demand for civil products would encourage a quicker reversion to pre-war conditions. The final choice was in effect even more radical with the 'earliest possible' being superseded by 'instant' for the discontinuation of war contracts because it was felt that it was 'undesirable that the output of useless munitions should be continued a day longer than was absolutely necessary'. The Committee's decision that such a principle could safely be applied was only made once it knew that 'unemployment allowances' would be paid to civilian war workers who were put out of work by the termination of government orders. The basis on which the Ministry of Munitions then made its plans was that reversion to peacetime production had primacy, and considerations of the subsequent effects on labour were not significant in its thinking.

It seems that so far as the optical industry was concerned, there was no knowledge of this policy, and the Ministry's own historian conceded that arrangements for demobilization were still not finalized when the armistice unexpectedly arrived. But no matter how incomplete were the plans overall, those for immediate contract terminations certainly were in place and they went into operation at once, greatly to the consternation of the optical munitions makers.

Contract Cancellations: Impact and Effects

The instant and sharp reactions not only show the optical munitions industry's ignorance of what had been planned, they clearly illustrate the extent of the optical industry's dependency on government orders and simultaneously the degree to which its products had become essential to the prosecution of the war. R. & J. Beck, by then one of the country's largest optical munitions contractors, wrote to OMGD on 13 November that the cancellation of their dial sight and trench periscope contracts would cause 'the immediate lay-off of 1,300 men'.[5] To avoid this, Beck asked for an arrangement similar to the first option considered by the Committee on Demobilization and Reconstruction, whereby the contracts could be run down rather than cut off short. Six days later, the Ross Optical Co., which shared dial sight contracts with Beck as well as having other large orders, told OMGD that the Contracts Department's instruction that all supplies were to be discontinued within three months meant that about half their 700 employees would be thrown out of work.[6] Like Beck, Ross asked for help in avoiding this. E. R. Watts & Co. (one of the many conscripts) said their entire working capital of £10,000 was tied up in optical munitions contracts, and they had no civilian orders at all. Adam Hilger Ltd, a much smaller firm whose importance in the industry was disproportionate to its size because of its specialization, warned that it would immediately have to make thirty redundancies and then progressively to dismiss every one of the optical glass workers taken on during the war, thus putting it back into very much the same condition that the OMGD's efforts had been meant to transform.[7] Having paid scant attention to its future during the war, for whatever reasons, the industry was now sharply aware of the wind of change blowing into its factories.

These contract terminations seem to have been made without any prior liaison whatsoever with OMGD, either because of the incomplete state of the Ministry's planning or because the Contracts Department had no instructions to make any exception to the general policy. The optical section, however, having worked hard towards improving the industry's overall condition in the previous three years, showed immediate concern over the potentially damaging effects of such sudden large-scale cancellations.[8] As early as 15 November, there were discussions with the Treasury about whether contracts could indeed be slowed down to allow a transition from war work and minimize redundancies.[9] Whilst insisting that it could only show preferment for a brief time, the Treasury conceded that there was actually no objection in principle as the optical industry was seen as a 'new one', although a decision needed to be taken quickly about what the subsequent course for its future would be. Three days later, the Minister of Munitions asked for a schedule of likely redundancies, to which OMGD replied that although it was not possible to produce an exact figure, the estimated total for the whole of the optical and scientific industries was about 10,000.[10] The Minister was assured that arrangements would be made to minimize losses, but there was no hint of what these would be.

There was ambivalence possibly amounting to duplicity in OMGD's response to the question of redundancies. On the one hand was the genuine desire to minimize the effects of cancellations on contractors, but on the other was the clear suggestion that the problem was of limited scale. If there were indeed only about 10,000 workers involved, then the problem might be regarded as relatively small by the government, with the implication that its solution might not be expensive or controversial. This was an echo of Frederick Cheshire's strategy in approaching the question of overhauling the industry in 1915 – the deliberate management and presentation of truth to create an image which would facilitate a desired outcome. A letter from R. & J. Beck to the Controller of Optical Munitions on 18 November said that the firm had been told to propose a scheme that would keep a reasonable number of its workers employed and avoid losing an efficient organization which might useful for rebuilding the optical instruments industry.[11] Beck's letter did not say by whom the instruction had been given, but a reasonable assumption is that it came from OMGD's Administrative Director, and was intended to support a case for holding together as many skilled workers as possible.

Whatever the sentiments of OMGD respecting the industry or the plight of its constituents, it was nevertheless apparent that the vast quantity of optical munitions on order for Britain and her allies was no longer needed and that almost all contracts would have to be ended prematurely. Although the industry was bound to be adversely affected, there were safeguards embedded in the contracts process which provided for payments as compensation when a contract was terminated by the state.[12] An Optical Munitions Liquidation Committee was quickly set up to manage the cancellations, and by the end of November it had delivered a report.[13] Its responsibilities were to decide which contracts to close, which, if any, to maintain, and to scrutinize the performance of companies which might be eligible for payments under the termination provisions. To qualify for compensation payments or 'liquidation amounts', a maker not only had to have suffered from a contract's premature ending, but also must not have been in default of its terms. A series of 'preliminary investigations' was started to assess whether, from the state's point of view, contactors had actually complied with the terms and conditions of the orders placed, so qualifying for settlements.[14]

This may have been a sound basis for good practice and due diligence, but because of the often confused placing and revision of orders, any rapid and accurate assessment of compliance was likely to be extremely difficult. That the OMGD fully appreciated the situation many of the contractors found themselves in is clear, perhaps not least because having striven hard to transform them under a policy of rejuvenating the whole optical industry, it had what might loosely be termed a vested interest in their survival and future prosperity. There was certainly an inclination towards supporting the makers in their appeals for financial assistance, although as will be shown, the OMGD was by no means blind to the failings of some of them. The Liquidation Committee, on the other hand, was looking for evidence of non-compliance, and had no brief to assist the contractors.

The first to be scrutinized in detail were the two largest London contractors, Beck and Ross. Other firms had been subjected to preliminary reports, but on 29 November the Liquidation Committee decided that further, urgent 'Special Investigations' should be held on Beck and Ross. Although no reason for this discrimination was given, the likely justification was because of their very large contracts for the expensive artillery Dial Sight No. 7, whose liquidation would involve very considerable sums of money.[15] The Controller of Optical Munitions advised the special investigators that the Ministry was contractually obligated to take an additional 4,550 sights at a cost of £230,000 – about £11.6 million in 2011 values – whether it wanted them or not.[16] After rangefinders, dial sight contracts had been the largest in value and were the Ministry's greatest single liability. They had been numerous, subject to much amendment and were a mixture of 'running' ones, which called for minimum delivery rates over an indefinite period as well as those for a specified quantity, which not infrequently had no delivery deadline at all.[17] For Beck and Ross, this was very valuable long-term business, and its premature ending must have been viewed with great concern. Such contracts' compensation provisions allowed for less than the full value of the contract, and a 'War Break Clause' allowed the Ministry to escape from the contract altogether under certain circumstances; evidence of default in particular would absolve the state from any liability for payment.

If the Ministry of Munitions was hoping that the Liquidation Committee would produce a verdict finding grounds for escaping payment through the contractors' default, then it must have been greatly disappointed by the findings of the special investigators. They reported in less than three weeks, not just on Beck and Ross but on all the other firms that had been subjected to 'preliminary investigations'. The Liquidator's subsequent report amounted to a clear admission of *nostra culpa*:

> practically none of our contractors have been able to maintain the contracted rates of delivery in view of the pressures put upon them by this department – investigation clearly shows that default has not been due to negligence or circumstances within the control of the contractor.[18]

It is tempting to think that when the special investigators went into the offices of those being scrutinized, they examined the sheaves of orders, amendments, alterations, cancellations and re-instatements held by the manufacturers, shook their heads in despair at what amounted to contractual chaos and then went back to write a diplomatically worded report that exonerated the trade whilst not specifically blaming any one party at the Ministry or the OMGD. To establish any case for default would have required a detailed and necessarily long investigation of each individual contract, far longer than the actual time taken to look over those investigated. The report simply pointed out what everyone intimately connected with optical munitions supply already knew – that the process of procurement and supply during the war had been characterized by pressure and confusion, had been immensely complicated by shortages of capacity, raw materials and labour, and not infrequently exacerbated by delays in acceptance by the War Office, often through squabbles over the minutiae of quality control.[19] To single out the contractors and penalize them financially would, in the majority of cases, have been unreasonable.

The report may have been made more acceptable, or even encouraged, because by then 'the labour situation was so difficult' that the Minister of Labour was asking Liquidation Officers 'to avoid any action which might result in violent dislocation'.[20] Arrangements for paying unemployment allowances were still not in place, and there was a belated recognition that, as the armistice was still no guarantee that hostilities were finally over, it would be imprudent to disperse the means of munitions production prematurely. But those fears were disappearing by the start of 1919, and the process of termination resumed after a brief but important check that brought some small measure of relief to the optical industry.

Having accepted that there were no general grounds to escape paying for prematurely terminated contracts, the Liquidator of Optical Munitions Contracts told the OMGD in early January that the general policy was to close down all contracts for 'stores not required by mid-March'.[21] The ending of losses by attrition when fighting stopped and the drastic reduction in the Army's size as men were demobilized meant the ending of practically all War Office orders as demands rapidly fell to pre-war levels. The policy after mid-March was to be one of ordering 'commercial articles' wherever possible in order to aid civilian industry. For the 'regular' optical munitions makers, this was hardly a blessing.

Although the War Office might support civilian industry by ordering off-the-shelf products, the amount of commercial articles the War Office would need for its optical inventory was virtually nil. Specialized optical munitions were now held in very large quantities and requirements would be minimal for the foreseeable future, so that the specialist capability of the firms recently drawn into optical munitions work was now redundant. It required little prescience to recognize that British military contracting in the foreseeable future would support none of the wartime conscript industry and by no means all of the regulars.

Although compensation payments had been virtually guaranteed by the Liquidation Committee's findings, the instructions to cease production meant that skilled operatives would be without work until new business was found. Employers were not just unwilling to pay an idle workforce, their limited and dwindling resources made it impossible. Firms urgently needed some means to keep trained workers productively employed whilst attempts were made to recover pre-war business and slowing down contracts would have helped to do this. In fact, the Controller of Munitions Supply had told OMGD on 16 December 1918 that although all contracts extending beyond eight weeks were to be terminated, he now had authority to slow down optical contracts rather than cut them off.[22] As his remit extended far beyond optics, this suggests that the industry's need for some special treatment might then have been recognized. However, on 10 January 1919, the Liquidator's announcement about closing down contracts showed that this policy had been revised and replaced by one of making cash payments instead, an alteration that seriously compromised earlier intentions to nurture a new industry.

The extremely dire situation of some firms is shown in surviving contract liquidation correspondence.[23] R. & J. Beck announced that it had no new designs for commercial markets because the technical design staff had been 'entirely engaged' on war work since 1915 and development of civil lines had therefore been precluded. E. R. Watts & Son had become totally dedicated to war production, had used up its financial reserves in extending its capacity for it and, now having no free capital, was 'financially embarrassed' with the loss of Ministry orders. The Ross Optical Co. reported on 20 January 1919 that all its optical munitions contracts had been cancelled and 'manufacture thereon has been stopped'. Adam Hilger Ltd was in such a state that the firm's directors had applied to the Ministry of Munitions for 'relief in respect of hardship through cancellation of contracts'. Dollond & Co. Ltd was 'totally engaged on government contracts' with over 4,000 prism binoculars in process of manufacture which were no longer required. These, the firm bewailed, had prominent government ownership marks that were 'impossible to remove', eliminating any chance of selling them commercially even if buyers could be found. W. Ottway & Co. Ltd summed up the seriousness of the generally depressing situation in mid-January, writing that since the armistice their efforts 'all over the place' to get orders for what they sold commercially before the war, had produced orders totalling scarcely £100. Hard times had indeed rapidly come on the optical industry. For its members, the twofold problem of surviving until business could be recovered, and scaling down both capacity and workforces from the levels generated by the war was exacerbated by a third. Overlaid onto the need for transition were financial difficulties rooted in the measures taken during 1915 to direct labour and control profits and wages.[24]

Despite the business it had received, the industry was by no means cash-rich at the end of 1918. Incomes had been high, but wartime profits had been geared to the levels of the last financial year before the war. Earnings had also been taxed at increasingly higher rates as the war went on, and the Excess Profit Duty levied on surpluses further severely eroded retained earnings. The production of accounts during the chaotic conditions after 1915 was generally delayed, meaning that provision had to be made during the war for still-uncertain but undoubtedly large amounts of taxation. Operating costs had risen steeply from 1914 levels and the increasing rate of inflation, particularly in 1917 and 1918, meant that many contracts had been far less profitable than expected when they were signed. Expensive additions to premises and plant had not always been funded by the Ministry of Munitions, and had been financed either from cash reserves or with borrowed money bearing interest. Although the state sought to limit contractors' profits, it did nothing to indemnify them against loss nor did it guarantee their liquidity.

The Ministry's surviving contract records indicate just how substantial was the scale of compensation payments.[25] Those surviving show that Beck was due to receive over £180,000, the Ross Optical Co. entitled to almost £83,000 and Sir Howard Grubb & Co. was owed £20,500 for parts and optical tools alone. E. R. Watts & Son asked for an advance of £6,500 to cover their immediate needs; this was paid immediately and without demur, suggesting that the total owing was substantially more.

There had also been many contracts placed independently by the Admiralty up to November 1918. Submarine periscopes, complex prismatic gun sights for warship turrets and naval rangefinders were never ordered through the Ministry, even after it formally assumed responsibility for Admiralty requirements in July 1917. Barr & Stroud's records show its Admiralty orders by 1918 had increased to a level that was considerably greater than War Office ones,[26] so it is reasonable to assume that Grubb, Ottway and Ross were also due substantial amounts for the naval element of their business. These contract liquidation sums represented palliative redundancy payments for what amounted to the closing down of almost the entirety of the expanded wartime optical munitions industry and the state's disengagement from any active policy of nurturing a broader optical instrument making community.

Specialization and Survival: Barr & Stroud, the War Office and the Admiralty

Given the extent of the difficulties being faced by the more broadly based firms, it might be expected that Barr & Stroud would have been even more seriously concerned by its prospects as it had no civil business to fall back on and the large-scale termination of contracts represented the loss of most of its work. The company was in a somewhat curious position in 1919. On the one hand, during the war it had extended its competencies and skills and become self-sufficient in virtually all aspects of the design and production of rangefinders, so it was potentially even stronger and more competitive than it had been in 1914. Expansion and vertical integration into glass production and optical computation had removed the two main pre-war weaknesses, and, in theory at least, equipped the firm to move into optical work outside munitions contracting. On the other hand, the greatly extended factory was set up entirely for ordnance manufacture and the company's expertise was wholly in making munitions-related instruments which were marketed through processes quite unlike civil products. Although, as already described, Barr & Stroud was largely insulated from many of the immediate problems facing almost all of the optical industry, the firm was by no means immune to the difficulties of adjusting to a peace that was very different to that of 1914.

The firm's treatment and its reactions to the war's sudden ending were quite unlike those of the others, most of whom had signalled immediate and pressing problems. On 15 November 1918, a meeting was held between Barr & Stroud's Harold Jackson, the OMGD's administrative head, Mr Knowles and a Captain Johnson, who was presumably from the Ministry's Contracts Department.[27] Jackson was asked to give 'a considered statement' on the question of rangefinder manufacture and uncompleted orders. His summary indicates the complexity and confusion surrounding optical munitions ordering. Only a month before, it had been agreed that an existing large rangefinder contract be reduced from 2,000 to 1,500 instruments, but it was then decided that another 500 were actually needed to cover the 'immediate needs' of Britain, Greece and the US. The first contract was already being delivered, and uncompleted instruments for it were in such an advanced state of assembly that Jackson stated it would be 'inadvisable' to terminate manufacture. As for the other 500, he thought that the Ministry should agree to their completion, work having already started and the firm's costing methods not providing for 'an accurate estimate of accounts' until the entire order was completed. Barr & Stroud's position was that the Ministry should pay in full for whatever it had ordered, but unlike other makers, there is no evidence of desperation or financial embarrassment. Three days after that meeting, Jackson indicated that the directors were happy to have the contract for a very large, and highly expensive, experimental coast-defence rangefinder cancelled, so long as the Ministry paid for all 'out of pocket expenses' incurred.[28] Unlike its English counterparts, the company gave every impression of being very much in control of its affairs, although lay-offs were already starting.

The first to go were the female workers that Barr & Stroud had begun recruiting once the need for dilution became pressing in 1917.[29] Despite the detailed planning done for their training and integration, it was always made clear that they were there only for the war's duration as the firm expected to re-employ those men who had earlier joined the armed forces. On 19 November Jackson warned the female workforce that redundancies were imminent and inevitable, advising those who wanted to remain in employment to seek alternative work before large numbers of other redundant women swamped the existing vacancies in the traditional areas of female employment.[30] The next day he informed the local Employment Exchange of the first of a series of redundancies in the male workforce which would begin on 30 November. The women numbered approximately 400 out of a total payroll of about 2,000, but there is no surviving record of the rate at which they were released. Presumably their dispersal was rapid, because by the year's end only 112 men out of 1,600 had been given notice,[31] which was a much smaller proportion of the workforce than the London firms had threatened to dismiss at short notice.

This relatively small number of redundancies would have been because Barr & Stroud still had a considerable amount of Admiralty rangefinder and submarine periscope work which was not subject to the threat of cancellation. Over £140,000 of orders were placed during the first ten months of 1918, besides work in progress from the previous year's total of £264,000, and other earlier still uncompleted contracts dating from 1916.[32] These were not affected in the same way as War Office orders, because they related to ship construction programmes rather than immediate issues to troops. Warships already fitting-out were in no danger of cancellation, even though the urgency of completion had disappeared, and would still need their outfits of rangefinders. Besides those, large numbers of anti-aircraft rangefinders were also on order for ships in commission as well as submarine periscopes for boats nearing completion. In addition, the Admiralty had ordered a number of experimental long-base rangefinders and mountings for trials in an effort to solve the gunnery problems disclosed at the Battle of Jutland in 1916.[33] Although difficult to assess precisely, the total value of Barr & Stroud's outstanding Admiralty business in November 1918 must have been substantially in excess of £500,000, with much of it looking safe from cancellation.[34]

There was also a considerable amount of War Office work, not all of which was expected to terminate prematurely. Before allowing for cancellations and consolidations, some £292,000 worth of orders had been placed by the Ministry between March and October 1918, excluding running contracts to repair rangefinders which had already been bought and used in service. Even if all the other

War Office business were to vanish, the repair work meant that a proportion of the workforce could be kept employed, at least in the short-term.[35] Scarcely a week after the armistice, the company decided to extend by a year the lease on its Woolwich Arsenal premises where instruments for repairs and routine overhauling were prepared for shipment to Glasgow, clearly signalling it had no immediate fears of its War Office business vanishing overnight.[36]

Nevertheless, there were still the problems of diminished government trade. The substantial rangefinder contract discussed in November, for example, was worth over £100,000 and its early termination involved a substantial loss of profit even after compensation. In the meantime, deliveries continued and there was the question of payment for them. Established practice was for payment being made only after goods had been inspected and accepted at Woolwich Arsenal, a procedure which had always been subject to delays.[37] The Ministry had been persuaded, or more likely bullied, by Harold Jackson into paying a monthly 'standard advance of £12,500' for instruments delivered and waiting inspection, but in January 1919 it proposed to terminate the arrangement.[38] The firm would have none of this, pointing out that dealing with the Ministry caused a 'very unsatisfactory' cash position, illustrated by a December debit balance to the Ministry of £37,921, plus another £50,000 for work in progress. Rather than terminating the advance, said Jackson, the Munitions' Accounts Office should increase it to £35,000. The Ministry would not accept Jackson's exhortation to be more generous, but it did abandon the notion of cutting off payments, maintaining the existing arrangement until the final liquidation settlement in May 1919.

Pressure on Barr & Stroud to accept cancellations on the Ministry's terms evidently grew. By the end of January, Jackson had refused to agree that no further deliveries would be accepted after certain dates, as well giving clear notice of Barr & Stroud's stance on the subject as a whole:

> We cannot ... accept cancellation of contracts without compensation, more especially as the reason for non-adherence to original delivery dates has arisen from causes entirely beyond our control. If therefore any of our contracts are deemed to be cancelled ... we reserve the right to claim full compensation.[39]

Jackson was quoting almost verbatim from the earlier Special Investigations report.

The skirmishing continued, particularly over the large order for 2,000 rangefinders. In early March, Jackson told the Liquidator that if the Ministry had not 'interfered' then 'the full number of instruments would have been delivered, and presumably the full profit would have been made'.[40] The 'fair way' was for the Ministry to pay for all materials and labour used, plus the previously agreed 'oncost' factor of 70 per cent, plus 10 per cent of that total for profit, and another 10 per cent on the grand total for 'royalty'. To show the firm's understanding of

the War Office's difficult situation about being saddled with unwanted equipment, Barr & Stroud would, if its proposals were accepted, buy the balance of the
undelivered rangefinders at 'an agreed price'. The firm was asking compensation
for some 650 instruments at the rate of £44.20 each, and its offer to buy them
back amounted to £5,790, or less than £9 per rangefinder.[41] Even though far less
than the cost price, it was still no trifling sum, implying either that Barr & Stroud
could easily afford the purchase, or that there was some client in mind for them.
If the latter was the reason, then it came to nought. No sale followed and many
of them were still in store in the late 1980s at the time the company moved to a
new factory when they were given gratis to any employee who wanted one.[42] By
29 January 1919, the Liquidator had agreed to almost all the company's terms,
except that he offered only 5 per cent for the royalty element.[43] Jackson stuck to
the firm's guns: 'We think we are entitled to some special consideration', he said,
as they had been 'inconvenienced' and in any case it was now 'very hard' to sell
rangefinders. As a parting shot, he reminded the Liquidator of a government
notice following the armistice saying that 'Contractors for war materials would
be liberally dealt with'.[44] By 2 May, the Liquidator had apparently given way, and
almost all the negotiations over War Office contracts were ended. Jackson then
wrote that £21,377 had been agreed upon for the rangefinders and 'Everything
is now settled I think, except for our claim for ... binocular prisms',[45] something
whose significance appears later in this chapter.

Although the Ministry had officially taken over responsibility for Admiralty
supplies on 30 June 1917, Barr & Stroud continued to deal directly with the
Navy over rangefinder contracts. Sorting out those cancellations was more protracted and difficult than for War Office ones, and only concluded in 1925 after
much negotiation and not a few threats on both sides. Discussions began with
the Director of Admiralty Contracts in August 1919 and displayed the firm's
willingness to exploit its de facto monopoly position.[46]

Prices for wartime Admiralty contracts had generally been calculated on the
same basis as for the Ministry of Munitions, the formula being 'labour costs + materials costs + oncost factor + profit + royalty'. During the war the profit and royalty
elements were standardized at 10 per cent each, with materials and labour being the
actual prices involved. The 'oncost factor', devised in the 1890s to cover overheads
and background expenses, was normally set at 70 per cent. The firm remained content with this formula when negotiating settlements for War Office contracts with
the Ministry, but things went differently with those for the Admiralty.

Contracts made in 1917 and 1918 had sometimes departed from the usual
arrangement, being made on the basis of a fixed sum to cover the oncost, profit
and royalty elements. In 1919, when these came under scrutiny for liquidation,
Barr & Stroud became concerned that compensation payments would not, from
the firm's point of view, fairly reflect the value of lost business. In August, Jack-

son pointed out that when made there had been the expectation of a long period of continuous work and that both wage and commodity costs had then been lower.[47] Subsequent rising costs had 'rendered quite inadequate the sums originally considered fair and reasonable to cover oncost plus profit' so that some fixed-price contracts were already showing a loss and others only the barest margin of profit, making it absolutely 'imperative' for the firm to ask for a revision of the prices.

The money involved was substantial, Jackson estimating in mid-September 1919 that the amount already owing on them was 'considerably over £50,000'. There were also a large number of contracts where even the basis for payment was still to be fixed. The company requested that all terminated contracts be settled on one basis irrespective of previous agreements, amounting in many cases to a retrospective increase in prices. To do this took no little nerve and showed considerable self-confidence because the fixed-sum contracts had been freely entered into, there was no legal ground for renegotiation, and by leaving open the matter of final profits the company had put itself in a weak position. As the client no longer urgently needed the instruments, Barr & Stroud's position might be considered to have been somewhat weak, an understanding clearly shown by the Contracts Department's initial response, which certainly showed no inclination to help. Its director peremptorily demanded a full list of the contracts involved as well as the full disclosure of profit and loss accounts from 1913 to 1918.[48] In addition, he stated that he was prepared to send the matter to the Treasury Contracts Committee which had the power to make a final decision against which there could be no appeal, a clear implication that by arguing its case the firm might consequently find itself even worse off. Quite unshaken by this, Jackson sent only the list of contracts, which, he pointed out, the Contracts Department already had. The demand for accounts was dismissed because 'We do not see that they can furnish any useful information'.[49] In fact, there were as yet no results for 1917 and 1918. Like other firms, Barr & Stroud had found that the war's chaotic trading conditions meant that accounts were not only difficult to prepare but largely meaningless without some means to disentangle the confusion generated by the involvement of the Ministry of Munitions in day-to-day business, as well as some mechanism to allow for inflation.[50] As for the threat to refer matters to the Treasury, he challenged the Director of Contracts to do that 'with the least possible delay'. Having thus lit the touchpaper, the firm then had to watch it burn – a slow process which continued for six years before the Admiralty eventually settled all the claims virtually in full, finally paying out £356,808.

These apparently fraught negotiations with the Contracts Department had no immediate effect on work in hand although, as will be seen in the following chapter, the whole relationship between the Royal Navy and the firm increasingly came under scrutiny as the Admiralty and the nature of that association began to change. Its new orders in 1919 totalled scarcely £10,000 but there was still a substantial amount of other work from earlier orders in progress, submarine periscope and anti-aircraft rangefinder development was proceeding steadily and a set of experimental rangefinders for trials nearing completion.[51] Besides all this, the company was examining and reporting on wartime German instruments as they came into the Navy's possession, so that both technical staff and production workers were being kept busy for the moment.[52]

Although prepared to fight for as much as possible from its cancelled contracts, and although certain that a reduced level of Admiralty business would continue, Barr & Stroud, like all the other demobilized optical munitions contractors, was faced with a pressing need to secure additional work to avoid further contraction. Some of this might come from foreign navies which had been starved of up-to-date instruments during the war, but in early 1919 this was uncertain and hedged about with political considerations as the recent combatants considered their post-war positions. Diversification into civilian markets was one way for Barr & Stroud to hold its plant and workforce together.

Previous writers have given both a 'broad-brush' picture of the firm's efforts at generally extending its product base into civil products during and after 1919, as well as examining in detail its attempts to manufacture and market binoculars.[53] These accounts suggest that the firm's subsequent limited success in these efforts resulted from a combination of inexperience in selling to non-government customers and a lack of adept direction. Such problems reputedly resulted in financial benefits that were at best minimal and sometimes non-existent. However, there is evidence which indicates that these attempts were never expected to let the business expand to any great extent through diversification, and that the concept of failure in commercial marketing is not really appropriate. Instead, there is a strong case to be made that Barr & Stroud never saw any substantial future for itself outside optical munitions manufacture and that the apparent attempts at diversification were really mechanisms for internally generating subsidies to support core activities in the lean times anticipated following the war's end.

The decision in 1919 to start production of binoculars has been singled out for not being propitiously chosen and indeed 'the worst period in history to launch such a venture', as large quantities of war-surplus glasses were coming onto the market as soldiers returned home with improperly retained Army-issued instruments and War Office surplus stocks were starting to be sold off.[54] Doubtless some servicemen held on to their binoculars, but whether there would have been enough to compromise any commercial market is a matter for debate. However, on the question of war-surplus sales, an official decision on the large-scale disposal of stores was not even made until 23 September 1919, when the intention put before the Cabinet was that

> All Government Stores in the UK, and in every theatre of war and all ports what-
> soever to be declared surplus forthwith and sold as soon as possible, excepting only
> sufficient to provide for the peacetime requirements of the Fighting Services and such
> duly authorised reserves as prudence may require in the interests of safety.[55]

The Government's desire was clearly to be rid of as much as possible in the short-
est possible time, without regard to prices fetched:

> The intention is to release storage and circulate stores and materials without delay
> and for this reason sales should be effected even at reduced prices, rather than hold
> out for better results which would entail the retention of storage accommodation
> which the commerce of this country so badly needs, and which is hindering trade.[56]

Although there was thus the possibility of binoculars being put onto the market
at very low prices in late 1919, this could hardly have been known by the company earlier in the year. And, given the continual shortages of such instruments
being complained about in the Ministry's own wartime weekly reports, the high
wartime attrition rates and the large quantities retained by the Admiralty, the
notion of an avalanche of very cheap binoculars swamping the domestic market after 1919 is perhaps difficult to sustain.[57] Other reasons to account for the
firm's decision include the expectation that the substantial pre-war German
export trade in optical goods would not be resumed, and partly through the
firm's 'restless ambition marinated in optimism'.[58] The first reason certainly has
merit, even if in early 1919 there had not actually been any indications that any
peace settlement would restrict pre-war German trading activities to the extent
of embargoing the manufacture or export of optical goods.[59] Certainly, the disruption of pre-war distribution and marketing structures suffered by German
firms in Britain meant at least a temporary lack of optical imports. The war
had completely closed down their pre-war presence and resulted in the loss of
premises, sales networks and marketing activities as well as the sequestration of
patents. The problems they faced subsequent to the armistice in first overcoming
anti-German sentiment and then resuming their trade contacts, recovering lost
assets and generally rebuilding their operations delayed the return of German-
made instruments to the British market.[60] Even if not excluded in the long term,
German optical companies were definitely absent for the present.

Far from being 'the worst period in history' to begin manufacturing binoculars, from Barr & Stroud's position the timing was not only good but even
imperative. The decision to start in early 1919 had nothing to do with ambition
and little with optimism, but was rather the result of the firm's realistic assessment of its immediate needs and opportunities, combined with a piece of sharp
commercial opportunism.

One of Barr & Stroud's immediate problems, perhaps the most important of
them, was the retention of its optical workers. Building up a self-sufficient opti-

cal department had been difficult because precision optical working in Glasgow was unknown before the company set up its own glass-working shop. Unlike the London-based optical companies, which were concentrated geographically and effectively provided themselves with a pool of skilled labour, Barr & Stroud had never been able to recruit experienced workers locally. Before the war the business either had to entice workers from far afield or train new ones from scratch. In 1919, for the first time Barr & Stroud had a full complement of optical workers carrying out all the processes needed to make even the most complex rangefinders. Losing them would put the business back to the unsatisfactory pre-war state when it depended on outside contractors who were frequently unable to work to the required standards. The only commercial optical product with sophisticated lens and prism systems similar to those of the rangefinder, and which had any prospect of being sold in substantial numbers, was the prismatic binocular.[61] It offered an excellent chance for keeping the experienced optical workforce employed in the face of uncertain prospects for further government orders.

During the war, Barr & Stroud had contracted to make 120,000 prisms for the Army's 'Binocular No. 3'.[62] This was the first time they had been mass produced in Britain, and the firm had acquired considerable expertise in making what was the most expensive optical component of the binocular (see Figure 7.1).

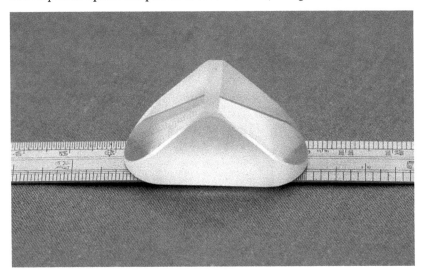

Figure 7.1: A binocular prism of the sort produced by Barr & Stroud Ltd. Approximately 1.5 in. long, four of these 'porro prisms' were needed in a binocular. Author's collection.

When the war ended, the contract was incomplete, and a large number of prisms were still at Glasgow in various stages of completion. The rough-moulded glass blocks for them had been supplied, at the Ministry's expense, by Chance Brothers Ltd for Barr & Stroud to grind and polish. When the prism contract was closed down, Barr & Stroud not only claimed compensation but also offered to buy the blocks still at Glasgow at well below cost, as well as tendering to buy a large number of binocular 'optical sets' (spherical lenses) from the Liquidator of Munitions Contracts.[63] With all the optical components thus to hand, the firm then needed only to provide the mechanical body parts and assemble everything. Binocular manufacture could thus be started quickly and then be done mainly using existing components which had been bought far cheaper than the normal costs of manufacturing.

The decision to start production was not to satisfy any 'restless ambition' or even to enhance the firm's profitability, but to safeguard the future of the optical workshop and its skilled staff; as one of the firm's directors later wrote, 'something had to be done immediately for the sake of the optical workers ... one of the objects of the decision was to keep at least some of the [optical shop] machines in operation'.[64] The binocular project was the product of what was, for Barr & Stroud, a typical combination of altruism and pragmatism. The desire to provide employment for the optical shop went beyond a philanthropic ideal in wanting to do something simply 'for the sake of the workers'. Disbanding a large proportion of the skilled optical hands would have severely weakened the ability to handle future munitions contracts and the venture was conceived principally as a way of holding together highly trained operatives who would subsequently have been almost impossible to replace. The directors, through a quick recognition of the potential utility of newly surplus prisms and lenses, set about creating what amounted to an internally generated subsidy to keep the optical shop in existence.

The company also began making 'cinematograph' machines to the design of a company which planned to sell them to movie theatre operators.[65] These cinema film projectors had some similarities with Barr & Stroud's munitions products. Besides an optical system, they used geared driving and other mechanical components which were related to the mechanisms in the rangefinders already being made at the factory. The firm's machine tools and its workers' skills could readily be applied to their production, and it was expected by the marketing company that demand for them would quickly begin to increase substantially.[66] The Ross Optical Co. and A. Kershaw & Sons Ltd, both optical munitions makers during the war, also made similar machines, and Kershaw in particular had considerable experience with them that dated back to before the war.[67] By April 1919, Barr & Stroud had a contract for 530 machines, and work began on them during June.[68]

The profits to be made were, however, substantially less than those from military and naval work. The contract had a gross value of £26,500, which the firm regarded as unprofitable compared to government work. Each machine sold at £50, a figure apparently based on the actual costs of labour and materials and significantly influenced by other makers' prices. This was a very different basis than the company's long-established formula for optical munitions work under which the price would have been at least double. Unlike optical munitions, the movie projector market was well supplied and competitive, with no great need for innovation. and technical superiority not particularly a telling sales point. In the cinema projection business, price was the overriding consideration. Business manager Harold Jackson believed that the return was inadequate in relation to the margins the firm had always enjoyed on its munitions contracting and was only prepared to accept it because he felt it was essential to keep the plant employed for the company's immediate welfare.[69] Although there might be some future profit, the principal motivation was again to retain as many skilled workers as possible. As with the binocular optics, even relatively rudimentary work at a reduced profit was better than losing highly experienced fitters whose value was largely in their savoir-faire.

The other moves towards diversification taken in 1919 were never likely to generate significant income.[70] Substantial profits were again less important than actually doing work in exchange for some income. A simple golf practice device known as the 'Impactor' was made, again for another company to sell, keeping some of the apprentices busy and allowing their retention at minimal cost to the business. It was only ever made in small numbers and, like the cinema projectors, eventually abandoned when its parent company failed. Another unprofitable exercise was the 'Optophone', a complex electrical instrument intended to convert printed words in books into sounds by which the blind would be able to 'read books by ear'. That was an actually an expression of Archibald Barr's philanthropic character but never proved truly successful in its operation and was eventually allowed to fade away, once again having shown a loss to the business.[71] A plan to manufacture motorcycle engines, which employed some of the mechanical techniques used in making rangefinders, was also taken up in 1919, but it developed slowly and the decision to start production was not taken until the end of 1920. Eventually, after both technical and marketing difficulties, it too was terminated without any financial benefit to the firm.

None of these moves into commercial lines generated much in the way of profits and emphasize the extent to which the company's expertise and success was connected with optical munitions. Diversification for Barr & Stroud was not necessarily principally intended to move into new territory, but was a means of holding together as much of the workforce as possible until munitions work established a new equilibrium. The months following the war's end were characterized by uncertainty about the future for the whole of the optical industry but Barr & Stroud alone of the optical munitions makers made the decision to remain wedded to that now unpredictable speciality. Where all the other firms saw their futures back in the civil market, Barr & Stroud banked on its core activity returning to a level great enough to sustain the business, recognizing that it lacked the expertise or infrastructure to move quickly and successfully into large-scale new activities which might substantially replace optical munitions work. Although accepting the need for short-term amendments to its product range, there was no doubt that rangefinders would continue as the mainstay of the business with the addition of submarine periscopes which had been successfully added to its stock-in-trade during the war.

The year 1919 was indeed significant for optical munitions manufacturing in Great Britain. It saw the large and specialized wartime industry vanish completely within a few months as the need for its existence disappeared. In one way, this was in accord with what the Ministry of Munitions had planned for. When Frederick Cheshire was its joint head, he had worked to cultivate a climate in which the whole of optical instrument making in Britain would be brought up to a level where sophisticated apparatus could be designed and made economically on a scale large enough to let the industry compete profitably with foreign makers in peacetime, and to meet all the country's needs in the event of war. The notion of a separate optical munitions industry had no place in Cheshire's philosophy, and much of his efforts had gone into creating what he saw as the essential underpinnings of scientific training which would benefit technical optics generally. That his ideas for reform were incomplete when the war ended was unfortunate and the sudden, premature casting loose of the industry from government work was bound to have serious implications for the future of the optical trade. The vicissitudes suffered by the makers of civil optics after 1919 are really separate to this account, and have yet to be examined and analysed. What is particularly relevant is that, for practical purposes, the optical munitions industry actually reverted to its pre-war state leaving only one substantial maker which now had to struggle to keep its capabilities intact. The problems encountered by Barr & Stroud were analogous to those suffered by specialist armaments makers, and heighten the case for considering optical munitions production as a distinctly separate activity to instrument making in general. The weakness of Cheshire's ideas lay in failing to recognize the commonalities with the bigger armaments industry and the similar problems of how the capacity for expandable production could be maintained in peacetime. The result was that the nation's capability for optical munitions production, after four years of trial and effort, was left in no better state than it had been in August 1914, with only one company retaining either the inclination or ability to stay in the game. Over the next four years, Barr & Stroud was to become, to all intents and purposes, the British optical munitions industry, as the following chapter relates.

8 ADAPTION AND SURVIVAL, 1919–23

Having passed through the upheaval of industrial demobilization, the optical munitions makers had to face the longer-term problems of the reversion to peace. The post-war period brought new political attitudes to armaments and the resulting shift to arms limitation and reduction contrasted sharply with the pre-war years, which had been characterized by the willingness of governments to spend heavily on military technologies. Budgets shrank and with demobilization the victorious armies found themselves with a surplus of optical munitions, many of which would not need replacing for a considerable time. This virtually eliminated short- and medium-term demand for land service instruments and for most pre-war producers military optics ceased to be viable business. Matters were different with sophisticated naval instruments such as large rangefinders and submarine periscopes, where demand survived because continuing evolution in related weapons technologies sustained the need for improvements, even though the quantities required were relatively small. These diminished requirements meant that by 1923 only one British company was still substantially involved in complex optical munitions production, and of the other makers mentioned previously, only three continued occasionally to manufacture less complex military optics. Frederick Cheshire's gloomy assessment of the capacity for optical munitions manufacture in 1915 perhaps came closer to the truth in 1923 than when he originally made it; even fewer firms were now involved and the capacity for mass production was far less than before the war. This chapter examines the policies and strategies for survival devised by the major participants and discusses the extent to which they were successful. It provides a reminder of how the optical industry was faced with a range of external factors which crossed social, economic and political dimensions and added to the internal problems relating to the technical aspects of instrument design and production.

The Makers' Problems

Optical munitions producers have, as described earlier in this volume, previously been seen as part of general instrument manufacturing rather than the armaments industry. It has been assumed therefore that their problems were those of the makers of civil optical goods, namely long-standing problems of inadequate research and development, difficulties of providing appropriate education and training, and the threat from foreign competition which benefited significantly from exchange rate advantages.[1] The optical instruments trade as a whole certainly had its share of such difficulties, but they were by no means the same as those facing the manufacturers of military or naval optics. The problems of optical munitions manufacture had a quite different grounding.

The principal difficulty for makers of complex service optics was not connected with scientific or industrial backwardness or inadequate business acumen or organization. Scientific training was not lacking at Barr & Stroud in Glasgow, for example, nor were research and development departments absent there or at Thomas Cooke's in York. The increasing complexity of the most important optical munitions had made them the province of a very small number of manufacturers who certainly were not without the necessary expertise to design and manufacture them, and because of the insistence of domestic supply by the armed forces, at least in peacetime, questions of foreign competition were generally irrelevant in sales to the British War Office and Admiralty. Rather, the difficulty after 1918 was that the massive reduction in post-war arms budgets made continued involvement difficult because of the paucity of orders. Adding to the problem was the peculiar nature of key products such as naval rangefinders and submarine periscopes compared to civil instruments like microscopes and spectrometers, which almost wholly precluded the possibility of spin-off that could have aided other commercial production.

Sophisticated munitions instruments needed technologies and manufacturing facilities unlike those normally employed in civil instrument making. Although precision devices requiring tolerances similar to laboratory instruments, large naval rangefinders and periscopes were massive objects requiring tools such as welding torches and 30-ft lathes in their manufacture, as well as heavy lifting gear to move them around workshops. Their lenses and prisms, although often weighing individually less than an ounce, were embedded in massively complex frameworks that frequently weighed nearly a ton and whose construction required specialized tooling and abilities that were unknown in, and frequently irrelevant to, the civil precision instruments industry. To move between one field and the other was not a straightforward exercise.

Because of falling demand and the trade's idiosyncratic nature, few firms showed the inclination and ability to remain substantially involved with optical munitions production. The speedy departure of the wartime conscripts to the industry has been described in the preceding chapter, and most of the pre-war participants which had run munitions work alongside other manufacturing largely wrote off government contracting as a regular, substantial activity. They returned to their civil products and continued substantially as before with just a handful occasionally making sighting telescopes or prism binoculars when the War Office placed small orders for them. Only A. Kershaw & Sons Ltd of Leeds and two London companies – W. Ottway & Co. Ltd and the Ross Optical Co. Ltd – received any War Office business up to the end of 1923.[2] They all regarded military contracting as only a very small part of their business, picking up and developing their pre-war activities as best they could. The small group of businesses which had been involved in the most specialized forms of optical munitions production faced greater problems, with a variety of outcomes which were not always beneficial to them. Sir Howard Grubb & Co. never resumed munitions activities after the November 1918 armistice and eventually went into liquidation, whilst Thomas Cooke & Sons Ltd struggled for several years to remain active before also being wound up. Barr & Stroud, whose specialized business had always lacked commercial lines, found it easier to remain in optical munitions than to diversify into civil products. The differing degrees of success of these companies illustrate how difficult it was to survive, let alone make headway, with optical munitions manufacture in the early 1920s.

Sir Howard Grubb & Co.: A Major Casualty

The one important optical munitions manufacturer that failed entirely to make the transition back to its pre-war status was the original maker of submarine periscopes, Sir Howard Grubb & Co. of Dublin. The firm had made them right from the introduction of submarines in the Royal Navy in 1901 and had been the sole British maker until 1911 when, at the request of the Admiralty, Kelvin, Bottomley & Baird of Glasgow acquired licences to manufacture the designs of the German Goerz company and the Italian firm Officine Galileo.[3] The company shared some of the monopoly characteristics of Barr & Stroud, but with significant differences that had an important bearing on its ultimate fate. In 1917, the company had been forced by the Admiralty to shift periscope production from its long-established works at Rathmines near Dublin to St Albans in Hertfordshire, supposedly to reduce the risks from U-boats when shipping completed instruments across the Irish Sea and partly through security fears over political unrest in Ireland.[4] The move had not been sought by Grubb and the choice of the new site was not the firm's but had been imposed by the Admiralty, which had then seemingly left the firm to organize itself as best it could, despite the urgent need for the instruments involved in the relocation. Perhaps in consequence of these factors, the relocation was slow and still far from complete when the war ended, leaving the business in a difficult situation from which it never recovered.

Grubb's post-war problems were worse than for any other long-established firm recently engaged in optical munitions. The disruption of the move had interfered not only with Admiralty periscope production but also with War

Office contracts for less complex instruments such as artillery sighting telescopes. Because deliveries had been greatly in arrears for reasons outside the Ministry of Munitions' control, it had been possible to terminate most of its War Office contracts after November 1918 without paying the substantial compensation received by most other contractors.[5] The interpretation of performance clauses to the Ministry's benefit, despite much of the difficult situation having been forced on the company by the Admiralty, denied the business the useful injections of cash that helped many others to ride out the difficulties of translating their manufacturing back to pre-war commercial activities. As late as March 1919, although work on small instruments was still being carried out in Dublin, the St Albans factory remained incomplete and had yet to start production.[6] This was not the result of any flawed policy of the company, and it is hard not to feel some sympathy for Sir Howard's subsequent plight because the move had been forced on him through the Admiralty's powers under the Munitions of War Act. If possible he would have preferred to abandon the project once the war ended, but the exercise had developed to the point where by November 1918 it was impossible to reverse matters.[7] To make things worse for Grubb, early in 1919 neither the Admiralty nor the Ministry of Munitions were eager to relinquish the wartime controls they still held over the business. The Dublin works were under military guard to protect government stores from expected civil unrest, and at the time Grubb believed his operation would not be free of state interference in the near future. Coupled with the Dublin works' partially dismantled condition and the still unfinished state of the St Albans site, this meant that he could neither take up unfinished pre-war contracts for astronomical telescopes nor complete whatever government work there was on hand.[8] Adding to these woes was widespread industrial unrest in Ireland which both prevented the shipping of large amounts of essential machinery still missing from the new works and interfered with operations at the already disrupted Irish factory. As far as can now be seen, the Admiralty subsequently abandoned the firm to its own devices and the Ministry of Munitions washed its hands of the affair as soon as the winding-up of its contracts was completed. There was little or no enthusiasm for the whole project within the company right from the start and, with some prescience, Sir Howard noted in 1919 that 'I am afraid we are in for a bad time'.[9]

He was quite right. By August 1921, even with the transfer to St Albans finally complete, things were still not going well. The recession had set in, and Grubb was in low spirits; he wrote to an overseas client who was still awaiting a large and expensive astronomical telescope ordered before the war that it was very difficult to convey to him how bad the state of business generally was, with factories closing or closed all around and virtually nobody placing orders.[10] His business was by then very much in straitened circumstances. Periscope manufacture, the mainstay of Grubb's optical munitions work, had stopped completely once the move from Dublin began and never again restarted. Almost no new civil astro-telescope work had come in, existing orders were very much behind schedule and, in consequence, the business was barely surviving.

The loss of the periscope business was critical and contributed substantially to the firm's dire condition in the early 1920s. Grubb, although by inclination and experience an astronomical telescope maker, had been drawn into optical munitions work around the time of the Boer War through an association with the armaments firm of Vickers, and the production of submarine periscopes had helped the business financially before 1914.[11] Although astronomical telescopes were high-value artefacts, they were never renowned for producing high profits, and payment for them was frequently protracted.[12] Grubb may have used income from his periscopes to subsidize telescope production; he had certainly benefited from the regular substantial payments that the early manufacture and servicing of periscopes had brought in.[13] Although he had enjoyed a monopoly of supply for almost a decade, its nature was significantly different to Barr & Stroud's position with rangefinders. Unlike Barr & Stroud, Grubb had never dealt directly with the Admiralty, either individually or as a business, and consequently had lacked opportunities to establish the same rapport that Barr & Stroud enjoyed. His relationship with Vickers, established when they first became involved with submarines, evolved into that of subcontractor, and his periscopes were supplied direct to Vickers (for many years the monopoly builder of submarines in Britain), who delivered the boats fitted with them. The Navy's technical requirements for periscopes went first to Vickers, probably because they were seen as part of the vessel's structure. Vickers then passed them to Grubb. In consequence of this separation, Sir Howard was unable to build up the connections that Barr & Stroud created with the Royal Navy, and this ultimately contributed in part to his loss of monopoly periscope manufacture.

A lack of contacts, though, was by no means the only reason for Grubb's failure to retain dominance as a supplier. The performance of the periscopes designed and manufactured by Grubb had been criticized by the Royal Navy as early as 1911 and the Admiralty's Director of Naval Construction had then looked at what were thought to be potentially superior designs from French, German and Italian optical companies.[14] Despite its policy of insisting on British manufacture for service equipment, the Admiralty bought trial periscopes from all three foreign firms and then encouraged the Glasgow-based engineering company of Kelvin, Bottomley & Baird to take out licences for domestic manufacture of the German Goerz and Italian Officine Galileo patterns. This ended Grubb's status as sole British periscope maker and should have acted as a warning that a monopoly or even the continuation of orders was not necessarily to be counted on. The arrival of competition apparently encouraged Grubb to improve significantly the mechanical design of his periscopes, but his weakened grip on the Royal Navy's business was further loosened in 1915 when the Admiralty approached Barr & Stroud to see if a rangefinder could be incorporated into existing patterns of periscopes. According to the company, such a fusion was impossible but the firm offered to integrate a rangefinder into a completely new design which could readily replace the instruments in boats already in service.[15] The proposal was accepted and a contract placed in 1917 not only for it but also trial models of conventional types similar to those being made by Grubb.[16]

This involvement of Barr & Stroud in periscope design and manufacture came at a most inopportune time for Grubb. Apart from coping with the inevitable problems of the Admiralty's demand that periscope manufacture be shifted to England, Grubb's business was ill-equipped to provide what the Admiralty was asking for in 1917. His company not only lacked any expertise in rangefinder design to create new periscopes of the type now being required, it was also without a research and development structure that could have attacked the problem vigorously. There would have been little opportunity to compete for extra business even had the transfer been completed satisfactorily. Barr & Stroud not only had a large design department of proven ability and a well-equipped and secure factory but, equally importantly, it had come to command the confidence of the Admiralty. Whether the Navy's decision to draw Barr & Stroud into periscope manufacture represented a disposition to marginalize Grubb is a matter of conjecture, but the deliberate disruption of the country's largest periscope maker without the prospect of an alternative supplier would have been a foolhardy step. The only other producer, Kelvin, Bottomley & Baird in Glasgow, were mechanical rather than optical engineers who must have relied on outside suppliers (possibly even from their neighbours, Barr & Stroud) for their glass components.[17] Whatever the underlying thinking at the Admiralty, this combination of circumstances finally and completely removed Grubb's primacy in submarine periscope making in Britain and transferred that role to Barr & Stroud, which not only replaced Grubb as the Navy's principal supplier but eventually attained a monopoly role.

Grubb's company struggled unsuccessfully after 1918, and the St Albans works never became fully operational. In 1920 he tried to reorganize for astro-telescope work on an expanded scale, but this was never carried through.[18] By October 1922, the company was said to be weak financially and probably surviving only through the goodwill of its bankers,[19] and in 1923 there were still large amounts of redundant material in the new factory including parts for submarine periscopes whose contracts had been terminated in 1919.[20] The relocation of the business had a catastrophic effect on its ability and efficiency; it ceased entirely to produce optical munitions, continued to decline as a builder of large astro-telescopes and eventually, in January 1925, went into liquidation.[21]

Grubb's failure as an optical munitions maker resulted principally from four mutually reinforcing factors. Added to the company's compulsory relocation to England was the underlying lack of capability to develop radically new devices and the Admiralty's consequent readiness to bring Barr & Stroud into periscope manufacture which then provided an opportunity for that firm to usurp Grubb's previous role. The fourth factor was Grubb's limitations as a business manager. Not only did he lack the power or the negotiating skills to dissuade either the Admiralty or the Ministry of Munitions from the idea of relocation, but he was also unable to resolve the resulting chaos once the war ended. The cessation of income from munitions business and the failure to secure compensation payments for terminated contracts emphasized Grubb's financial weakness and seriously compromised his ability to take up his pre-war commercial work with astronomical telescopes. It also crippled his prospects of continuing to be competitive with his conventional periscopes which, in the event, were not immediately superseded and continued in service for many years afterwards.[22] For Sir Howard Grubb, the war had brought neither short-term nor lasting benefits. His business, more than any other in optical munitions production, suffered through serving the state.

Thomas Cooke & Sons Ltd: Another Casualty

The eventual fate of what had become Cooke, Troughton & Simms Ltd was very much the same as Sir Howard Grubb's business, but its route there was very different. Cooke's had a broader base in optical work, including service optics, than Grubb, although its pre-1914 importance to the British state was actually far less. The company had ceased to be independent when Vickers acquired 70 per cent of its shares in 1915, a circumstance that bore heavily on its post-war course. The support of the massive armaments company should have placed the business in a stronger position to compete in optical munitions work, but Cooke's struggled to make headway after the war, losing money until eventually its parent withdrew support and, like Grubb, it went into liquidation. As with Grubb, the company lacked any close rapport with the Royal Navy, and despite the strong connections its parent company already had with the Departments of Naval Ordnance and Naval Construction, it never built up the kind of relationship that had always worked in Barr & Stroud's favour. Nor did the firm develop much of a relationship with the War Office before 1914, despite its involvement in producing earlier types of rangefinders and its loose pre-war connections with Vickers.

The relationship between Vickers and Cooke's is not always easy to understand, and even the reason for buying such a large stake in the firm is unclear. The firm's own historian considered it was to secure control of just one particular product, by which he could only have meant the mechanical fire control instruments that Cooke's were already supplying to Vickers for installation in warships.[23] This may be correct, but Vickers was probably at least as interested in the connections Cooke's had with the Argo Co. Ltd, which was the marketing agency for the Pollen system of naval gunnery control that had been rejected by the Admiralty in 1914.[24] Not only did Cooke's make all its mechanical elements, the firm had also introduced a sophisticated long-base rangefinder to be used in conjunction with it that was radically different from the Barr & Stroud pattern and promised to be superior in certain respects.[25] Cooke's had a majority interest in Argo, so that by acquiring the one, Vickers acquired the other as well. Cooke's was the only other British company besides Barr & Stroud to have conducted systematic rangefinder research before the war and had constructed a number of large naval rangefinder prototypes. The business was highly competent in optical design and offered Vickers an opportunity to integrate optical munitions capability into its other armaments operations. Cooke's importance to Vickers lay both in its optical expertise and its ability to tackle the specialized optical and mechanical engineering needed for both complex optical munitions and their related control systems. What Vickers did not do, though, was to take over the day to day running of the firm, nor even to establish a strong presence on its board.[26] Instead, the business was left very much to its own devices, with its munitions department coming under Cooke's general management and accounting structure, which unfortunately seems to have lacked the expertise of its technical division.

Although there was much to commend the idea of integrating Cooke's into Vickers' armaments operation as a base for optical munitions manufacture, the exercise ultimately proved costly and problematical. Cooke's had indeed developed advanced rangefinders during the war, however it had not been asked to make sophisticated optics for either the War Office or Admiralty, all their orders being for relatively simple sighting telescopes. Nor, prior to the war, had it actually succeeded in selling any rangefinders to either British or foreign governments because all the deals that were in progress in 1914 were frozen by the outbreak of hostilities.[27] As a result, after the war Vickers found that it owned an optical instruments company which, despite its technological potential, still had no proven record in complex armaments optics or any programme to develop new products, either civil or military, and which was also in poor financial condition. Unlike Barr & Stroud, Cooke's had no experience in selling to foreign armed forces, and its name lacked sufficient cachet to command the attention of prospective military buyers. All the potential for optical munitions sales actually lay in the firm's ability to design and manufacture instruments nominated by Vickers as a result of its own armaments experience. The eventual failure by Vickers' management to identify viable new opportunities for optical munitions designs meant that the parent company found itself propping up an instruments business that struggled to overcome the problems of post-war re-adjustment, whilst the few munitions products that were developed under Vickers' aegis failed to find markets either at home or abroad.

Cooke's had never been a particularly profitable business and although its accounting records have not survived, by late 1922 the company was clearly struggling despite its recent acquisition of one of its instrument-making rivals, Troughton & Simms Ltd of London.[28] This had actually been done with Vickers' encouragement and money, but it is unclear how Vickers might have expected to benefit from it.[29] Troughton & Simms was a smaller, long-established firm which had little expertise in optical munitions; its wartime activities had been on a lesser scale than Cooke's and its design capabilities virtually non-existent. It has been suggested that Vickers were persuaded into the acquisition because Troughton & Simms' management was better than Cooke's at producing competitive goods at relatively low prices.[30] This seems doubtful, particularly as the London firm was also financially weak and after 1920 suffered from a crucial lack of effective direction and management, with strong disagreement between the two principal family shareholders whose chief desire was to be rid of the responsibility of running the business in increasingly difficult conditions.[31] Whatever the underlying reasons for the purchase, the result was a further weakening of Cooke's overall condition and the merged company continued to decline.

In the early 1920s, the merged company of Cooke, Troughton & Simms was not actually selling any optical munitions although it had prepared designs and built prototypes of instruments that Vickers hoped subsequently to market. Vickers' aviation interests led to a 'Prismatic Bomb sight' being manufactured in 1919, as well as an 'Aeroplane Periscope' that allowed a pilot to see the ground directly under his aircraft.[32] The following year, Vickers' naval interests led to new designs of large naval rangefinders for Admiralty trials to decide on future standard types, and in 1920 and 1921 a number of prototype rangefinders and range-and-height finders were constructed for both surface and anti-aircraft use.[33] Although substantial business would have resulted if the prototype patterns had been adopted, their construction must have been an additional drain on Cooke's resources. The Admiralty was accustomed to having trial instruments supplied gratis, and the costs of developing them were expected to be borne by the submitting company, which may explain why the Vickers board was starting to register concerns over Cooke's finances and why it was prepared to keep underwriting the growing losses.[34]

The Vickers finance committee had forecast in September 1920 that Cooke's bank overdraft would exceed £58,000 by the following June. Cooke's bankers (who were not the same as Vickers') were uneasy about its financial situation, having recently refused to raise the existing overdraft ceiling of £45,000, and in consequence Vickers had to lend Cooke's £15,000 to meet immediate needs. In addition, another £75,000 was needed to cover liabilities for Excess Profits

Duty. By February 1921 Vickers' loans stood at £39,000, and Cooke's bankers were asking for the overdraft to be guaranteed, a further indication of their doubts over the firm's finances. The deterioration continued with a trading loss of £10,000 for the financial year ended 30 September 1922. The now-guaranteed overdraft's ceiling was increased with Vickers' support, first to £60,000 and then to £80,000 in the following January. The trading loss for 1922–3 worsened to £16,183 with unpaid Excess Profits Duty of over £22,000 still to find, by which time it was clear that Cooke's was no longer a viable business.[35] Vickers finally wearied of pouring money into a chronically ill business late in 1923 and liquidated the firm in the spring of 1924.[36]

The only optical munitions designs recorded at York from 1921, when Cooke's position was seriously worsening, until the close of 1923 were some observation periscopes for naval gun turrets and a prismatic sight for the Vickers–Berthier light machine gun which Vickers hoped to sell to the War Office and the Indian Army.[37] Like the earlier rangefinders and aeroplane instruments, these came to nothing. The observation periscopes, for some unexplained reason, were actually made by Barr & Stroud,[38] and the machine gun sight languished because neither prospective client made its mind up about the Vickers–Berthier gun. The rangefinders produced for the Admiralty trials in 1921 failed to convince the Royal Navy of their superiority, although Barr & Stroud certainly saw them as a serious threat and worried about prospects until May 1922,[39] and the aeroplane instruments apparently never went into production. Not a single Cooke–Vickers optical project successfully generated orders between 1919 and the end of 1923.

Vickers' attempts to capitalize on the integration of optical capacity into its armaments business did not fail because of the difficulties Cooke's had in marketing its civil products. The lack of success was first because the amount of new munitions business available at home and abroad was small and second because all the advantages lay with the established and demonstrably successful competition. Even if Cooke's had been a profitable instruments company, Vickers's efforts at developing a range of optical munitions would, almost certainly, still not have succeeded by 1923. That failure was almost inevitable, given the scarcity of domestic and foreign government business and Cooke's previous lack of success in selling ordnance products in competition with rivals. Faced with a dominant and proven domestic competitor, and in the absence of either demonstrable functional failure in the Royal Navy's existing rangefinding instruments or being able to demonstrate any inherent superiority in its own, Vickers had little chance of displacing Barr & Stroud as the Admiralty's preferred supplier. Cooke's failure as an instrument maker and Vickers' failure as an optical munitions supplier were quite separate issues.

Left alone, Cooke's would almost certainly not have continued with optical munitions after the war. Its earlier efforts had brought no financial rewards, and the firm probably felt it had suffered through its association with the vexatious Arthur Pollen's dealings with the Admiralty.[40] Wartime munitions profits had been small as a result of state taxation policies and like almost every other optical maker Cooke's finished in a weaker state than when it began as a munitions conscript. That the firm was a reluctant optical munitions player in 1919 does not necessarily mean it would have survived solely as an instrument maker. The vicissitudes of the British instruments industry were felt as keenly by Cooke's as by anyone else. The company's potential utility to Vickers justified continued support only until it eventually became clear that no substantial optical munitions work was likely in the foreseeable future, and that matters relating to Cooke's civil manufacturing could not be allowed to drift further. Any hope of immediate and substantial profits from civil instruments in 1920 was long gone by 1923 and it then made sense for Vickers to let the ailing company go to the liquidator, apparently relinquishing all involvement with optical munitions. However, Cooke's story did not end there; Vickers bought the company's assets and refloated it under their own direct control, so that they kept some capacity for fine mechanical and optical engineering. Under a new management they also ran the instruments business on a more or less profitable footing until the re-armament programmes of the late 1930s pulled Cooke's once again into optical munitions contracting.[41] For Cooke, Troughton & Simms Ltd, 1923 marked not so much the end of optical munitions manufacture, but the start of a period of hibernation following reincarnation.

Barr & Stroud: The Survivor

Unlike Grubb and Cooke, Troughton & Simms, Barr & Stroud not only stayed in business beyond the early 1920s but remained almost entirely dedicated to optical munitions. The company had the benefits of a continuing domestic monopoly of rangefinders and a growing presence in submarine periscope manufacture, as well as the valuable assets of a modern, well-equipped factory backed by a large research and development section operating under an effective and tightly controlled management structure. To these was added the determination to continue exploiting its previous success. However, success is a relative term, and in a post-war munitions environment Barr & Stroud sometimes found that it amounted to little more than simply remaining in business. Despite its corporate assets, the problems of maintaining the firm's existence were often considerable and not infrequently outside its directors' control. One critically important – though intangible – asset that Barr & Stroud possessed, uniquely amongst British optical munitions makers, was the symbiotic relationship with the Admiralty that had continued unbroken since 1892. This association, which brought both benefits and obligations to each party, was a very significant factor in Barr & Stroud's survival in the difficult times of the early 1920s.

The relationship was, nevertheless, not something that could be taken for granted by the company. The war had imposed strains on it, and in particular on the firm's earlier ability to set prices for Admiralty work as it alone thought fit. That had been overturned by the provisions of the Munitions of War Act which had imposed a series of controls intended to bring war work firmly under the state's governance both by allocating contracts and regulating profits.[42] Almost immediately after the Armistice, Barr & Stroud sought to be rid of those controls, causing the Admiralty to look closely at how it saw the future nature of their post-war dealings. The firm wanted to be able to negotiate freely over prices, to pursue whatever markets seemed opportune, and to be rid of the assorted wartime controls that, like Sir Howard Grubb, it saw as restrictive and irksome.[43] This request to be released had an effect on the Admiralty which Barr & Stroud probably could not have anticipated and which apparently remained unknown to the company at the time.[44]

Before responding to the request to be freed from the controls of wartime legislation, the Admiralty set out to consider the merits of control from the Royal Navy's viewpoint and looked for opinions within the Service as to the future of dealings with the firm. The responses illustrate that, despite the undeniable fact of an established relationship, there was by no means unanimity about how the company was regarded within the Service.[45]

The Naval Contracts Department favoured keeping Barr & Stroud permanently under Admiralty control not only by retaining the wartime arrangements but by extending them still further. It saw great advantages to the Service in keeping the firm's skills and technical facilities available on demand, although difficulties were anticipated in arranging what amounted to a takeover. State business would not be sufficient to keep the firm going, and government ownership would preclude not only foreign work but also other commercial production because 'the Private Trade would not appreciate Government competition'. A subsidy would therefore be needed, which the Director of Contracts recognized as being problematical to arrange. Nevertheless, he favoured bringing the company into the Navy's hierarchical structure, chiefly to control prices and avoid the contractual arguments which had arisen during the war.

Those arguments were not welcomed by the Director of Naval Ordnance (DNO) who saw Barr & Stroud in quite a different light. His response illustrates the Service's internal tensions about what should be the company's future status.[46] The DNO said that because the firm was the only British maker of naval rangefinders it was 'imperative' to keep it going both for peacetime needs and future requirements in time of war. That would better be done by maintaining the status quo. The idea of outright control disturbed the Ordnance Department because it threatened the satisfactory nature of the relationship that had grown up with the company; in effect, his department was perfectly happy to have an independent Barr & Stroud acting in perpetuity as its consultant, providing research and development work on what was represented as a no-cost basis. As with ordnance, where the commercial development of weapons had long been accepted as working to the Navy's advantage, so the design and supply of optical munitions was established as an external and indispensible adjunct. The firms supplying both had become so closely identified with the Navy's own interest that they were seen by the Ordnance Department as a now inseparable part of its functional framework. It strongly resisted any change in the relationship, seeking to maintain the familiar and mutually satisfactory arrangement that had evolved in the preceding twenty-five years, no doubt encouraged to do so by the complete absence of any viable alternative supplier.

Even worse than unwelcome organizational change, though, was the chance that the firm might fail completely. For the Ordnance Department the solution was to let the firm once again sustain itself with foreign business. The Admiralty's hierarchical structure, here represented by the Ordnance Department, was resisting change, not through conservatism or prejudice but through the reasonable and justifiable fear of a future failure in the provision of design and supply. The DNO insisted state control was neither 'advisable nor necessary' and urged that no other restrictions should be imposed on the company beyond those already in place, stressing that it would take until well into 1920 to complete current orders. In his judgement there was 'no other firm in the country who can be compared with Barr & Stroud in respect of their experience and facilities', and he urged an immediate meeting to settle what work should be regarded as 'specially confidential'. That would clear the way for Barr & Stroud to seek new foreign business and ensure its survival by making up any shortfall in Admiralty orders. In the end, the DNO's arguments won the day and the idea of perpetuating control was abandoned by the end of February 1919, probably without the firm ever having any inkling of it.[47] The advantages in preserving the status quo, where the Navy obtained rangefinder research seemingly free of charge, were massively in the Admiralty's favour, and the Contracts Department's sentiments were finally disregarded.

However, the wartime controls were not immediately rescinded, a situation that Barr & Stroud was temporarily, if reluctantly, obliged to accept.

The question of finding new business was indeed pressing. Contracts were disappearing and work to replace them was urgently needed. Foreign orders were most likely to come from the navies, both of the late allies and neutral powers, which had been starved of rangefinder deliveries since 1914, but some of the wartime controls, particularly those relating to secrecy, stood firmly in the way of opening such business. In a studied approach to the Admiralty, the firm's general manager, Harold Jackson, first assured their Lordships that 'nothing can give us greater satisfaction' than continuing to work for the Royal Navy, and that the firm would continue all the security precautions mandated by the circumstances of wartime, including not soliciting foreign sales without specific consent. Then, almost certainly with clients already waiting, he blandly asked whether, without further special applications, he could supply France, Italy, Japan and the US with 'any instruments actually in use by H. M. Fleet' at the date of the armistice. Jackson presumably got his way, because there were no further letters from him protesting about obstacles, but the problem of foreign sales for the new instruments such as extremely large rangefinders then being developed for the Royal Navy was less easily solved. He agreed to defer the matter until questions of limiting the spread of armaments 'attaching to the proposed League of Nations have been formulated and agreed by the Powers'.

There were now constraints on business that were unknown in 1914.[48] They went far beyond those imposed by the Admiralty and were likely to be an even greater problem. Jackson touched on the nature of future difficulties when he mentioned the limiting of armaments and the embryonic League of Nations, implying that Barr & Stroud already understood that future opportunities were likely to be governed by factors completely outside the firm's control. Much of its pre-war prosperity had come from supplying Europe's large conscript armies, none of whom were now in the market for instruments as forces of the victorious powers shrank rapidly leaving enormous surpluses of optical munitions. The neutral states which had been denied deliveries after late 1914 generally maintained only small armies and, even if they actually offered some opportunity for business, the value of their likely trade was small, especially as large numbers of rangefinders were now sitting idly awaiting disposal in the victors' depots at whatever prices could be obtained for them.

If the prospects of foreign land-service business were largely discouraging in early 1919, there was, despite the problems, rather greater optimism over naval orders. Even the Allied navies had been starved of rangefinders since 1914, and war experience emphasizing the need for effective gunnery control systems suggested there was a reasonable expectation of foreign business when navies sought to modernize their equipment. The firm must have had unplaced orders pending when Jackson wrote to the Admiralty in March that year, and there were actually still some rangefinders earmarked for foreign governments held in store from 1914 when deliveries had been embargoed.[49] As an antidote for excess optimism, however, there was virtually no new foreign warship construction starting outside Japan and the US. New ships had always been the prime movers for high-value orders, because with rangefinders went their expensive outfits of associated mountings and data transmission systems which the firm also supplied. The US Navy had instituted a large programme of capital-ship building, but Barr & Stroud had never captured its business as it had done with France or Japan before 1914, and after 1915 the British government had prevented the firm pursuing

American sales at all. By 1919, the US optical company Bausch & Lomb had advanced so far in rangefinder expertise (much of it derived from making Barr & Stroud designs for the US Army) that it was unlikely that Barr & Stroud could win major US orders, but Japan still lacked any sophisticated optical industry and the Imperial Navy had so many earlier ties with the company that Barr & Stroud must have seen it as the main chance of high-value foreign sales.

Although some later observers considered that the prospects for the short term seemed poor in 1919, with the directors pessimistically believing that there was little chance of any revival in the optical munitions trade,[50] this was by no means the case. The company was undoubtedly still busy. Turnover for the year was a substantial £369,279 with a workforce of 1,200, roughly the pre-war number, still employed in the autumn working a 47-hour week and – most importantly – an order book that was by no means empty.[51] The value of those uncompleted orders is not easy to assess as the firm's financial records are some-times hard to interpret, but the sums involved were certainly very considerable, probably exceeding £500,000, even if the future of some of the contracts was uncertain. Ordnance work was certainly not about to evaporate and the firm's board of directors was definitely not in despair. In fact, by November 1919, the directors had agreed unanimously that they could not even contemplate moving away from the armament business and that it must be maintained if at all pos-sible.[52] As for them being resigned to the prospect of no new munitions orders, in October Jackson asked the Admiralty to confirm there was no objection to quoting the Imperial Japanese Navy for new rangefinders, and began nego-tiations with the Coventry Ordnance Co. for the provision of a 'complete fire control system'.[53] In November the Admiralty asked for a design for a new range-finder for large submarines, as well as additional fire-control instruments for the torpedo-directing rangefinders on large surface ships.[54] And in December nego-tiations began with the Dutch army for 600 infantry and artillery rangefinders to make up for the dearth of deliveries since 1914.[55] Although these negotiations were all in the early stages, prospects were by no means lacking and confidence certainly not absent.

Even if prospects were far from barren, there was still the problem of manag-ing the present, because inflation and reduced margins had greatly eroded the firm's profitability.[56] The turnover of £369,279 was actually some 15 per cent lower than in 1913 after allowing for the heavy wartime inflation.[57] Manufac-turing profit was down from 55 per cent to 26 per cent, which with higher operating costs resulted in a small recorded pre-tax loss of £530. Bank overdrafts were £62,402, the highest since the limited company's creation. Against this, the year's new orders received were only just over £39,000, which meant that the value of work on the books was declining. Overall, the situation was far from ideal, although still a long way from crisis. However, 1920 saw a marked deterioration in the firm's position. Recorded turnover at current values declined to £310,822, manufacturing profits fell to only 12 per cent, and the year-end showed a very large pre-tax loss of £80,497. Although orders had increased to £91,114, this was insufficient to sustain the business, and after allowing for inflation, new work was only one-eighth of that received in 1913. Borrowings had increased to £129,497 by the end of December, and it appeared that without some radical change of circumstances the business would be heading towards insolvency. Faced with the biggest financial problems since its inception, the firm sought to achieve an upturn in its prospects through looking for a major reformation in its relationship with the Admiralty.

In the autumn of 1920, Barr & Stroud requested an annual subsidy. This marked a major change in the way the firm saw its standing with the Admiralty and prompted the latter again to reconsider its own role in the association. The company asked for £50,000 yearly in order to maintain experimental and research work and keep both their own factory and plant in good order, as well as buildings and tooling provided by the Admiralty during the war.[58] In essence, the company wanted the state to guarantee the costs of running the business, a circumstance inconceivable in 1914.

There can be no doubt that Barr & Stroud's position was far from satisfactory. This stemmed partly from diminishing business and partly from the twin burdens of maintaining an expensive research and development section whilst carrying a substantial amount of Admiralty debt. The research facility, which worked almost entirely on Admiralty projects, was the firm's largest standing charge and its salary costs had risen from 5.23 per cent of turnover in 1913 to 18.05 per cent in 1919.[59] Manufacturing wages fluctuated with output, but the cost of scientific staff had continued to grow irrespective of current production, and sustaining research and design was Barr & Stroud's heaviest single expense. A guarantee of £50,000 would cover it and many of the company's other standing costs as well. However, such a subsidy would not help with the question of outstanding bills still owing to the firm.

There were substantial Admiralty debts for wartime contracts where Barr & Stroud had financed the work's progress. Their total is now difficult to assess, and it appears to have been a problem even at the time. In October 1919, Jackson reckoned that the sum owing for finished and invoiced jobs alone was in excess of £58,600.[60] To complicate matters, invoices had not been submitted for many completed contracts because the basis for charging was still not agreed. There were seventy-three of these waiting to be settled in August 1919, whose value was not recorded.[61] To make matters worse, there were frequently long delays in payments for pre-priced contracts which forced the firm to press hard for settlement on several occasions, more than once even telegraphing requests for money.[62] Because materials and labour costs were paid by Barr & Stroud as contracts progressed, the burden of financing Admiralty orders continued to grow, even as the total value of work on hand was falling. At the beginning of 1920, these pressures were showing. Jackson was in discussions with the Department of Naval Ordnance about a large project involving the development of a new 30-ft rangefinder intended to become standard for capital ships, and on 22 January he wrote that unless relations with the Contracts Department improved the firm would ask for its name to be removed from the Admiralty list of approved contractors – adding that without the changes asked, its presence might be automatically removed by proceedings in the Bankruptcy Court.[63]

This was an illustration of the company's willingness and ability to play one section of the state's establishment against another. The need to obtain payment from the Contracts Department was paramount, but it was the Ordnance Department that stood to suffer from the possible loss of Barr & Stroud and it was to them that Jackson made both complaint and threat. The threat may have been delivered with a light touch, and a delayed action fuse, but the complaint about the Contracts Department's shortcomings was in earnest as many of the firm's immediate difficulties could be laid at its door. What the business needed was not only orders, but also prompt payment for them. In early 1920, Barr & Stroud's greatest difficulty was to match income to outgoings, and having to borrow large sums at interest to finance the Admiralty was not just expensive but clearly frustrating as well.[64] Jackson's complaint was part of his continuing efforts to extract the money due from the Admiralty, although by the autumn of 1920, those efforts had seemingly had little effect.

Faced with the subsidy request implying that the company was in difficulties, the Admiralty once again looked at the relationship. This time it passed the question to the Director of Scientific Research (DSR) for an opinion on the cost of the Navy itself maintaining a research department to handle rangefinder design and construction, as well as optical glass research.[65] The question was not whether the sum requested was reasonable, but whether the research could be done cheaper by the Service itself. The DSR made a general examination of rangefinder procurement and his report illustrated the position the Admiralty thought itself to be in vis-à-vis the company. Its content strongly suggests that, over time, Barr & Stroud had succeeded in colouring the Admiralty's perceptions very much to its own benefit. The DSR accepted that Barr & Stroud was in financial difficulties and that the consequences of its failure would be serious. There were no substantial new British orders likely for some time to come and he believed the firm's foreign business would only last for about two years. Without assistance, he thought Barr & Stroud would have to convert to commercial instrument making and close the rangefinder business down in order to handle the new work. That would put both the fighting services in 'a most dangerous position' as there was no other firm to step in. 'We must have rangefinders', he said, neatly summing up the Admiralty's predicament and begging the question of where else they might be obtained.[66]

If Barr & Stroud discontinued rangefinder operations, the creation of an Admiralty research and experimental department would be essential because no other firm would be in the position to make rangefinders without external assistance. The DSR estimated its annual costs at between £17,500 and £22,500 for research into just the mechanical aspects of rangefinder design, and suggested either setting up a government factory to produce entire instruments, or a state-owned assembly shop to assemble parts made by other instrument firms. His preferred plan was for the Admiralty to take over the large new premises it had built adjacent to the firm's Glasgow factory during the war and then progressively transfer both plant and personnel from Barr & Stroud as the company dropped out of rangefinder manufacture. This would avoid any 'dangerous hiatus during which the supply of rangefinders would be completely stopped'. As an alternative afterthought, he suggested that the state might buy a controlling interest in Barr & Stroud, which would let the Admiralty 'run the firm in the way they wanted'.

The DSR was unclear about what should actually be done, but an opinion dated 16 December 1920 from the Admiralty Research Laboratory at Teddington was much more decisive. This opinion acknowledged that the question of subsidy was problematical, but insisted that in the present circumstances there was simply 'no choice'. There was no existing state rangefinder factory and because it would take several years to create an effective one, 'some agreement with [Barr & Stroud] must, in the public interest, be arrived at'.[67] This was very much to the point and in line with the Admiralty's eventual conclusion which was that a subsidy would be the most effective and simplest way to assure the supply of rangefinders.

To help secure the necessary funding from the Treasury, the Admiralty then approached the War Office for its cooperation in the proposal. The War Office, perhaps mindful of its unsuccessful pre-war attempts to obtain rangefinders from other sources, agreed to the overtures, and the Director of Army Contracts subsequently recorded that

> in view of the probable smallness of orders in the near future, the Admiralty and War Office have conjointly made application for Treasury sanction to subsidize Barr & Stroud to enable that firm to keep in being its existing facilities for manufacturing.[68]

The paucity of current War Office orders could hardly be disputed, just £75 having been spent with the firm in the financial year 1920–1, out of a total outlay on optics of only £2,347.[69] Despite the conjoined overture, the application was vetoed by the Treasury which refused to provide funds because in its judgement the current warship building programme would provide enough work to keep the firm going.[70] Exactly how this conclusion was arrived at is hard to see given the absence of major warships then being built in Britain, but the decision stood, leaving the Services facing the prospect of Barr & Stroud withdrawing from rangefinder building. However, the course of subsequent events turned out to be very much different from what might have been expected.

When the Director of Army Contracts wrote his annual report in March 1921, he noted the Treasury's refusal but observed that it had not been possible to take any further action on Barr & Stroud's behalf, as the firm had made no further appeal for assistance.[71] Given that scarcely six months earlier the company had been predicting great difficulty in staying in optical munitions work, the lack of subsequent calls for succour must raise questions as to what had happened in the meantime. The year 1920 had continued to be difficult and in mid-September, Barr & Stroud reminded the Naval Staff's Director of Gunnery that although the Navy was calling for lots of new designs, there was still 'no real [i.e. paying] work' coming in.[72] Admiralty orders for new instruments and servicing contracts for the year came to just over £17,888, less than 20 per cent of the firm's new equipment business.[73] Without Japanese orders totalling £71,953, the year would have been catastrophic. However, the start of 1921 marked a sudden upturn in the company's attitude and its fortunes. January's orders came to £37,900, which apart from 1915, was the largest ever for that month.[74] In late February the bank overdraft was down from almost £100,000 at the end of December to only £54,000, 'without any Excess Profits duty repayment' and 'two or three large accounts still to be paid' to the company.[75]

The firm's attitude to armaments work in general and the international arms business in particular was neatly summed up by Harold Jackson's observation that Japan's recent decision not to accept any reduction of armaments 'may be sad from the humanitarian point of view, but it is not likely to cause much sorrow with Barr & Stroud Ltd'.[76] The Admiralty had pronounced the new 30-ft FX rangefinder model 'excellent', and was asking about an even larger one.[77] Armaments business was now starting to look more encouraging, unlike the civil ventures started soon after the war's end and mentioned in the preceding chapter, none of which seemed likely to be profitable. The 'Optophone' device was uncertain of making even a small return on its investment, the cinema projector programme was mired because the single client could not pay for the machines already delivered, and the motorcycle engine project was demanding such large sums that the directors had been obliged to talk 'solemnly about costs'.[78] Despite those difficulties, Jackson was able to tell a correspondent 'Don't think I'm not cheerful',[79] a sentiment which probably summed up the firm's overall attitude in early 1921.

There were indeed some changes for the better that year. Despite the prediction in late 1920 that the Royal Navy would have little business for Barr & Stroud in the near future, the Admiralty ordered almost £60,000 worth of equipment, making it the largest client in 1921.[80] The Imperial Japanese Navy was the next largest, with orders of nearly £53,000. The total value of new business that year was £125,610, an increase of 38 per cent on 1920's figure of £91,114. Although this was an encouraging trend, the state of new business was really not so much getting better as becoming less bad. Nevertheless, the underlying condition of the business was improving. Manufacturing profit increased to £98,629, up from 12 per cent to 36 per cent of output, and there was a small pre-tax profit of £4,406 compared to the previous loss of £80,497. Even more importantly, the accounts in December 1921 showed the borrowings at the end of 1920 had been discharged, and even after paying £26,000 of dividends there was still £10,765 cash in the bank, plus another £2,335 in French National Bonds. A fall in turn-over, from £310,822 to £259,226 represented the working-through of older contracts before payments for newer ones began, and continued the expected upward trend in sales.[81] Clearly, a major financial turnaround had taken place, but neither an increase in sales nor a reduction in operating costs could account for it. Although the surviving records fail to identify the reason, the most likely explanation was the receipt of delayed refunds for overpayments of Excess Profits Duty. These large wartime payments, totalling at least £225,000 according to Jackson's working papers, must have harmed liquidity, and their progressive, though undocumented, repayment would have been instrumental in restoring the balance sheet to a satisfactory condition.

The prospects for munitions business began to improve during 1921, enhanced by the Government's belated and reluctant decision in December 1920 to restart capital-ship building which had finally come to a halt with the completion of the battlecruiser *Hood* in May 1920.[82] No other capital ships had been planned after 1916, partly because the fleet action at Jutland that year had raised questions about what types of ships and armament were actually needed.[83] The post-war elimination of the German High Seas Fleet as a threat had been countered by the apparent willingness of the United States to complete its very large wartime construction programme that would have challenged the superiority of the Royal Navy and which was the subject of much contemporary political and naval debate.[84] In January 1920, the Admiralty had urged that four new ships be started in the financial year 1921–2, with four more the year after. In December 1920, the Committee for Imperial Defence agreed that whilst diplomatic efforts would be made to check the US's naval programme, the Admiralty could begin planning the ships it had advocated. Design studies had been progressing since 1919 so that plans for the first four ships were approved as early as August 1921, invitations to tender issued on 3 September, and orders placed on 26 October.[85] This must have been welcome news to Barr & Stroud. In March 1921 the Admiralty had approached the firm about constructing rangefinders with base lengths in excess of 40 ft,[86] and by June discussions were taking place with shipbuilders Armstrong Whitworth over the necessary turret installations.[87] The new ships were to reflect all the lessons of the war as well as recent progress in design, which required rangefinders that were larger, more sophisticated optically and more complex in construction to give more accurate readings. Not only was the firm's new 'duplex' design using two large instruments in a special mounting to be used for the main armament control, it was also to be used in a new, smaller, dual purpose distance and height-finding rangefinder for the secondary and tertiary armaments.[88] These were all vastly more expensive than those used during the First World War, and were to be provided on a much larger scale. The four battlecruisers were to carry at least three 41-ft instruments each for the main armament as well as sets of 15-ft duplex instruments for the other guns and torpedo armament, all requiring associated sighting telescopes, periscopes and fire-control equipment, representing a very substantial amount of business.

The optimism that may have been generated at Barr & Stroud by the ordering of these ships was to be of short duration. There had been considerable political reluctance to embark on a costly capital-ship programme in Britain, and a similar growing desire in the US to disengage from its own programme. In July 1921, the US had called a conference of major naval powers to discuss the whole question of naval armaments, and this took place in Washington during November of that year. The resulting Washington Treaty limited new building and fixed relative strengths between the navies of the signatories. Welcome as this may have been to the politicians, it had serious implications for Barr & Stroud because one of the key clauses was 'a ten-year capital ship building holiday'[89] which ended British plans to build the eight new warships for which the firm would have supplied all the optical fire-control apparatus. It also curtailed the Japanese programme which, in the continued absence of an adequate domestic optical industry, would have provided a significant amount of business for Barr & Stroud.[90] From anticipated orders for a programme of eight capital ships spread over a number of years, almost overnight the company saw its prospects for business reduced by three-quarters as the Royal Navy's building plans were reduced to just two smaller ships requiring a reduced complement of rangefinders and associated ancillaries.

This was a considerable problem for the firm, particularly in holding together its large design staff. Following what had become the normal pattern of Admiralty business, although none of the rangefinders for the projected battleships had actually been ordered when they were cancelled, design work was progressing and the firm's research and development department was fully employed on it. In the meantime there was not enough business in progress to maintain the large shop-floor workforce and by September 1921 the company was 'having a pretty thin time' and had been forced to lay off workers.[91] Despite this, there was no intention of letting go of any of the expensive research and development team, even in the absence of the subsidy which had been refused by the Treasury. Although the design department's costs accounted for most of the firm's salary payments – nearly all of the £29,347 paid out as salary in 1921 went to this department – it was recognized that it was both vital to the company's survival in the munitions business and its greatest resource. When the Admiralty Research Laboratory approached Barr & Stroud in December asking for help finding a skilled optical designer, the reply was that despite the adverse trading situation and difficulties in securing long-term business every effort had been made to keep such scarce specialists and not a single one had been discarded.[92]

This determination was in the context of an overall situation that whilst superficially improving was nevertheless fraught with problems. The post-war decline in sales, and the attendant losses, had been checked by the end of 1922, and the order book value had been growing since 1920, but the bulk of those orders were foreign ones and domestic contracts were insufficient to sustain the business. For instance, in 1922 Admiralty and War Office orders totalled only £55,000 in a total of £177,000, which even with a manufacturing profit of 52 per cent was scarcely able to cover even the firm's salary costs of £27,475.[93] Although this dependence on foreign business had been a familiar state of affairs before the First World War, there were now changing circumstances that put a very different complexion on the small proportion of orders from the British state. As recounted in the preceding chapter, Barr & Stroud had deliberately remained almost wholly dedicated to optical munitions work, so much so that in 1923 less than 4 per cent of its income came from products sold in the civil market. This in itself was a major achievement and a remarkable state of affairs, given that no other optical company at home or abroad had managed to do the same. No British firm was then producing military or naval optics of any type on a regular basis, whilst over 80 per cent of Barr & Stroud's output was made up of range-finders and almost the whole of the rest on other armament-related products.

The major German competitors, C. P. Goerz and Carl Zeiss, had, temporarily at least, been expelled from the field under the provisions of the Versailles Treaty, although Zeiss was seeking to avoid the proscription by setting up a company in Holland which would manufacture military and civil optical items as an ostensibly separate Dutch operation.[94] In the US, the Bausch & Lomb Co. had developed into a fully fledged producer during and immediately after the war, but ran its optical munitions business within the framework of a far larger business which was involved in ophthalmic and scientific instruments manufacturing. Although Bausch & Lomb had gained an effective monopoly with the US government and had established a clandestine arrangement with Zeiss, it had made little progress in foreign munitions sales outside the US.[95] Barr & Stroud, though, was still in a far from ideal position.

Only foreign sales made the business viable. British service orders for 1923 totalled £114,994, which on their own would not have sustained the business. Its viability came from foreign sales, 87 per cent of which that year came from the Imperial Japanese Navy, leaving Barr & Stroud largely dependent on it. Japan had always been the firm's largest overseas buyer of naval instruments, and in 1922 and 1923 had ordered rangefinders and commissioned a prototype mechanical analogue fire-control computer.[96] Japan still lacked an optical munitions industry with the ability to make large rangefinders as well as the high precision mechanical engineering capability necessary for the associated analogue computers needed to calculate a moving target's future position, but this was not something Barr & Stroud expected to continue indefinitely. For several years the Japanese Navy had had resident inspectors at the Glasgow works, and by mid-1922 the company knew that two of them had already returned home and were designing rangefinders and submarine periscopes intended to be built in Japan.[97] The firm was also well aware that Zeiss had set up a Dutch subsidiary to build optical munitions, and may have known from its resident Japanese contacts that Zeiss had already established connections with the Tokyo firm Nippon Kogaku as well as with Bausch & Lomb in the US.[98] German technicians had been seconded to the Japanese company in 1921 in order to facilitate both design and production and the anticipated growth in Japan's optical self-sufficiency may account for Barr & Stroud's willingness to continue its pre-war practice of sharing information about foreign navies' orders and interests with the Royal Navy.[99] The firm kept the Admiralty updated with details of the Imperial Navy's evolving interest in their gunnery control system for which the firm hoped to get a substantial order following the successful demonstration of the prototype in July 1923.[100] A serious setback to Japan's development of its optical munitions capability came with the earthquake of September 1923 when the Nippon Kogaku works were destroyed, causing delays on its route to self-sufficiency and benefiting Barr & Stroud by prolonging the Imperial Japanese Navy's connection with the company.[101]

Domestic orders in 1923 included part of the rangefinder outfits for the two battleships being built as a result of the Washington Treaty. Only the smaller instruments for the secondary armament were ordered during 1923, at a cost of £24,820, with the bulk of naval orders that year coming through £72,000 worth of contracts for similar rangefinders intended for cruisers. The very large main-armament outfits, worth £40,813, were not officially ordered until early in 1924, although work on them had begun during 1923.[102] These were significant because they were the last orders from the Admiralty for such enormous and complex rangefinders until the later 1930s, and effectively marked the suspension of this activity for almost fifteen years until the resumption of capital-ship building.

Barr & Stroud's survival as a specialized maker was simultaneously fortuitous and predictable. In refusing to divest itself of its munitions industry characteristics in 1919 and refraining from large-scale re-investment into civilian products of uncertain prospects, the company held on to all its skilled research and development organization and the greater part of its skilled workforce. Most of its forays into civil products were as much to do with providing work related to the skills of its shop-floor operatives as to opening up new avenues for business. The firm's directors reasoned that the market for its main speciality product, the naval rangefinder, would not evaporate and that its unique combination of specialized design ability and production capacity would make it virtually indispensible to the British state. In this they were substantially correct, even though attempts to obtain a direct subsidy were unsuccessful. The combination of unique technological abilities and the company's sustained symbiotic relationship with the Admiralty, coupled with the enduring characteristics of Britain as a 'warfare state', meant that the firm's survival was always more likely than its demise.

Into Suspense

By 1923, the British optical munitions industry was quantitatively, though certainly not qualitatively, inferior to its condition in 1914, with only one business actively engaged in producing instruments on a significant scale. This was not the result of inadequacies in technological ability, nor any lack of business acumen, but came about because the demand for military and naval equipment had fallen internationally to a level where all armaments producers were hard pressed to find adequately remunerative markets. Optical munitions makers fared little different to those making weapons after the war and were generally forced to diversify in order to survive. Barr & Stroud alone successfully maintained its status as a speciality producer, albeit on a smaller scale than before 1914. Capacity had adjusted to current demand, and the industry was at the start of what can be described as a period of hibernation during which the ability to produce all kinds of optical munitions would be sustained, even though output remained at a low level for the remainder of the 1920s and the early part of the 1930s. The disposition to limit armaments constricted the British optical munitions industry but failed to bring its elimination, leaving enough of a nucleus to build on when the shift to re-armament eventually arrived.

CONCLUSION

The story of optical munitions manufacture in Britain between 1888 and 1923 has shown the evolution of a peculiar, even idiosyncratic industry whose progress was often as much governed by the state's defence policies as by its technical and commercial abilities. Although sharing many of the characteristics of the optical instruments industry, such as the computation of lens systems and the finest standards of mechanical engineering, its evolution is best understood by seeing it in the perspective of the armaments industry which provided the stimulus for the creation of almost all of its products. The crucial division between civil instrument and munitions manufacture lay not just in the nature of the products but in the nature of their clients. Armies and navies, and their exchequers, were very different to surveyors, scientists and the man in the street requiring disparate marketing methods. But despite its differences, the optical munitions sector still drew on the private sector for many of its optical techniques, raw materials and skilled workers.

The British optical munitions makers functioned in a demand-led market which was driven by evolving weapons technologies and often heavily influenced by social factors exerted by those who formulated and influenced thinking within the armed forces. The War Office's protracted indecision in choosing an artillery rangefinder between the end of the Boer War and 1914 was caused at first by the service's adherence to the concept that the greatest potential accuracy in an instrument should take precedence over the need for speed and convenience of operation. Even when that concept was discarded, a final choice was delayed by extended comparisons between what were essentially two variations on a common theme. In the same period the Admiralty delayed adopting the larger and potentially more accurate rangefinders that were being bought by foreign navies, not because it could not see their benefits but because its gunnery branch was still arguing over the potential merits of two rival gunnery control systems. In both cases weapons technology had arrived at a point where the need for new optical devices was clear, but in neither case was the industry able to exert sufficient pressure to resolve the Services' indecision, even though the products were already available.

Although such difficulties were irksome, the industry was hardly held back by them. At a time when parts of British technological industry are often shown as fitting well with notions of relative decline, this one showed a pattern of financial and technological growth right up to the start of the First World War. It was aided by a climate of lavish and increasing spending on armaments in Europe and beyond, as well as domestically. In that politically and financially heated environment the British optical munitions industry became increasingly, and eventually totally, dominated by one company.

From its inception, Barr & Stroud deliberately chose to specialize in the rangefinder, which quickly became the most significant and remunerative type of optical munitions. The rangefinder was in many ways the ideal product for specialization. Its complexity and high entry costs for any would-be competitors helped keep new players out of the market and its high selling price made it a substantial revenue generator. In essence it was needed in two forms; armies wanted large quantities of the smaller models and navies required fewer of the much larger and disproportionately more expensive variants. With the latter went associated mounting and data-transmission systems which greatly added to its revenue generating capacity. The combination of specialization, close patent protection and shrewd marketing let the firm create for itself a dominant place in the world market for rangefinders. Indeed, it would be hard to find another British company in the armaments field that did quite so well. In foreign sales, it would be misleading to say that Barr & Stroud competed successfully against its rivals from a German optical industry commonly acknowledged as the world's largest and most accomplished, because the reality is that the German companies were forced to compete, almost always unsuccessfully, against Barr & Stroud.

Archibald Barr & William Stroud became 'first-movers' in rangefinder manufacture and established an early lead, not because they recognized a marketing opportunity and exploited it – as William Armstrong did with his breech loading artillery weapon after 1854 – but because they were serendipitously drawn as academics to a military problem that awaited a solution, and because of their willingness to persevere in finding an answer, even though the commercial rewards were not immediately quantifiable. Had they not been discouraged by the slow progress of their earlier joint academic research project in 1888, they would never have taken up rangefinder design, and had they been entrepreneurs by profession they would almost certainly have been so deterred by the difficulties they met in their rangefinder experiments that they would have abandoned the idea after their failure in the 1889 trials. Unlike Armstrong who first identified a market and then, stimulated by motives that were professed to be as much patriotic as business-like, invented a novel product through the application of engineering techniques that were familiar to him, the Professors Barr and Stroud stumbled accidentally across rangefinders and taught themselves as they went along. The rapid progress up their learning curve came from the application of scientific methodology which enabled them to become first-movers in the field and establish a momentum that let them build up a lead which competitors found extremely difficult to overcome before the outbreak of war in 1914.

Barr & Stroud's performance as a business can be interpreted in two ways. One, the upside, is that it was a success, running counter to notions of a general underperformance in British technological industries before 1914 and exploiting the benefits that first-mover status conveyed to secure both market share and profitability. The other, the downside, is that weaknesses in management caused a failure adequately to develop and employ the strategies of vertical integration that were needed to ensure the business obtained all the facilities it required along its chain of production. That lack can be said to have inhibited its development and limited its capacity for diversification out of a highly specialized and narrow market. Both of these interpretations contain elements of truth. The company was indeed profitable, and it undeniably had a hegemonical market position in 1914, but at the same time it did fail to integrate fully into lens and prism manufacture, a condition whose causes and implications must be understood before a judgement can be reached.

It has been suggested that the most common reason why firms should carry out vertical integration is to guarantee a constant and adequate supply of the materials needed for their production processes. That need certainly applied to Barr & Stroud, and its problems obtaining adequate quantities of high-quality optical components were manifest right from its formation, tending to worsen as the level of business grew more rapidly after 1912. The difficulty was exacerbated after the outbreak of war, reaching a level in its first two years when the inability to procure optical materials and parts at times threatened to halt production altogether. That failure, which might also be described as limited success insofar as no breakdown in output ever took place, can be said to have resulted from two possible causes. Established ties and personal relationships, typified by those shown towards Adam Hilger & Co. Ltd over the erratic quality of optical components, concentrated attention on improving the relationship rather than replacing it with another, more satisfactory, one. It is indeed true that Barr & Stroud sustained a repeatedly problematic association for an extended period, at least partly because of a longstanding relationship, but that was only part of the reason.

Even before the war, the company was prepared to make the necessary capital investment to set up its own optical department but was restrained from doing so by factors that were almost wholly beyond its control. Those factors reflected both the size and condition of British optical manufacturing in general, and in particular the lack of structures for scientific education and technical training outside London that had already been identified by the instruments industry itself. Geography, as much as anything else, inhibited the progression to optical self-sufficiency. The optical instruments industry was heavily concentrated in London and of the handful of provincial firms the nearest was some 200 miles away. It was not that Barr & Stroud did not want, or could not afford, to make such a move. The problem was that the only means to obtain the necessary skilled labour for producing optical components was from other optical businesses, all of whom were geographically remote. Once that impasse was ended by changes created during the First World War at the instigation of the Ministry of Munitions' Optical Munitions and Glassware Department then the firm moved quickly into large-scale optical production and integrated even further into optical glass design and manufacture.

The apparent success of Barr & Stroud combined with its uniquely extensive surviving archive tends to skew attention towards it and emphasizes an unfortunate – though unavoidable – reduction of attention on the other pre-war optical munitions makers, most of whom have left little in the way of records. For these other less well-documented players, it would be damning with faint praise if all that can now be said is that collectively they contributed to meeting the requirements of the British armed forces adequately, providing a source of supply that, contrary to previous suggestions, did make Britain independent of foreign makers after the Boer War. The greater importance of those manufacturers was that they provided a base on which to build much of the massively expanded wartime industry and that many of them not only survived the post-war retrenchment, but went on to perform a similar role in the Second World War.

The surviving records of the Ministry of Munitions provide an archive that, even if incomplete, allows the opportunity to judge how well the optical munitions industry responded to the challenges of the First World War. What remains provides a large body of data about what happened from mid-1915 to late 1918. Unfortunately, the printed *History* reflects only what the Ministry chose to record of what it saw as its main achievements, and the remaining manuscript material was heavily weeded out in an apparently arbitrary manner during the closing down of the Ministry in the early 1920s. Nevertheless, much unpublished material that was marked 'confidential' and never meant to be available to the manufacturers remains to show the 'coupled agendas' that ran within the Ministry's Optical Section. They represented largely unofficial efforts to inject state aid, not into optical munitions manufacture but into the peacetime framework of the civil instruments industry in an effort to bring it to a level of parity with its German counterpart. What emerges is a picture showing the diversion of the short-term energies needed to complete industrial mobilization and accelerate war output into a longer-term effort to create a strong, science- and technology-oriented instruments industry that could quickly and effectively adapt to large-scale munitions production in the event of a future war. That philosophy reflected governmental pre-war attitudes to the structure of armaments production which had envisaged private manufacturers supplementing the state arsenals to achieve adequate output, but in the case of the optical industry the notion ignored the essential difference that there was no state-owned capacity for manufacture. Cheshire's plan to secure future optical munitions supplies through a rejuvenated civil industry also contained the flaw of making no provision for keeping that sector of activity alive in peacetime when demand might be minimal.

One inference from that, and a very hard one to resist, is that despite what he had seen since the late spring of 1915, Cheshire was still failing to grasp the fundamental differences between almost all optical munitions and civil instruments in mid-1916 when his efforts to create a new infrastructure for advanced education in optics were well under way. Most of his plans were rooted in the entrenched pre-war attitudes and desires of the instrument-making community which he knew well; those sentiments had become virtually an intellectual paradigm and built up enough momentum to keep attention firmly focused on the need to create an optical industry closely modelled on the German one that had long been regarded as intrinsically superior. Neither Cheshire nor his colleagues recognized that, so far as optical munitions were concerned, there was no critical inferiority in the British model. None of that means that his emphasis on technical training was misplaced; to the contrary, it was the lack of it before the war that had prevented Barr & Stroud from successfully integrating into lens and prism manufacture and its provision was long overdue. His misjudgement was in failing to understand how very different were the vast majority of the instruments and the circumstances surrounding their marketing and sale.

It was ironic that Cheshire's success in pushing for university-level training in optical design should lead to him leaving the Ministry in 1917, before the rest of his ideas had matured to a point where results were likely to be forthcoming. His departure caused much of the momentum he had built up to dissipate, and little was done to set up the regional network of technical-school teaching that he had envisaged. The end of the war, which seems to have come much sooner than anyone in the Optical Section expected, brought all those developments to a halt and created chaos throughout the wartime optical munitions makers.

If the mobilization of the optical industry as a whole was largely successful, its demobilization went far less well from its members' point of view. The war's abrupt termination found the industry fully adapted to war work after a lengthy process of industrial mobilization, and both unprepared and ill-equipped to abandon munitions work and resume making civil products. All Alfred Esslemont's wartime energies had been used to create capacity for war products, and the cessation of new orders together with the almost immediate and total cancellation of existing government contracts meant that the wartime industry quickly found itself redundant and disbanded, having to resurrect civil product lines and seek out old clients as best it could. Of the pre-war optical munitions makers, only Barr & Stroud had been a total specialist, and it was the specialization that had brought its earlier prosperity that now threatened it with collapse. Barr &

Stroud never had commercial products and almost all its old foreign clients were either fully stocked with instruments or their market closed down by the war. The return to peace was a bigger trial than the war itself.

The problems Barr & Stroud faced in 1919 would almost certainly have occurred at some point, if to a lesser extent, even had there been no conflict. By 1914, most of the benefits it had enjoyed as a first-mover were being weakened, not because it had fallen behind in the technology of what it produced nor because of its inadequate integration in optical production, but because many of its foreign clients were either starting to move towards self-sufficiency in optical manufacture or were almost at that point. The firm's profits since 1905 had increasingly come from overseas armed forces in countries which either lacked the ability to produce complex optical devices or were prevented from making particular types, such as the rangefinder, through international patent protection. In 1912, for instance, Austria–Hungary and Russia did not yet have advanced optical industries and Barr & Stroud had started negotiations to supply both countries with rangefinders, key parts of whose designs were still covered by patents. Two years later, as the war began, technology transfer via foreign firms who had established factories there meant that both countries could already consider the production of sophisticated instruments, a condition encouraged by the approaching end of the patents protecting the designs.

The same applied in France and Italy, as well as the US, and the only other country likely to continue to be a large buyer of military rangefinders was Japan, whose optical industry was still relatively backward. But even there, prospects were limited to the mid-term rather than longer. Having long had technical staff seconded to Barr & Stroud's factory to test and adjust the naval rangefinders they were buying, the Japanese Navy had already accrued a body of expertise that would be the nucleus for a similar industry in Japan. The momentum of Barr & Stroud's success was starting to dissipate, not because of superior competition, but through a combination of growing foreign self-sufficiency, market saturation and the absence of any demonstrably superior rangefinder to replace expensive instruments whose service lives were likely to last for a goodly number of years. In 1914 the firm was reaching a condition that, but for the war, would soon have demanded attention to the question of alternative products.

For Barr & Stroud, then, the problem after 1918 was how to convert to peace in order to survive. The development stimulated by the war meant that, unlike in 1914, the firm was now largely self-sufficient in everything except large quantities of the simpler optical glasses, and could undertake to manufacture every item of optical munitions required either by the Army or the Royal Navy. It was the antithesis of Cheshire's prescription for a wartime optical munitions industry; not a civil business converting on demand to military products but one that made only optical devices for warfare and lacked the ability to manufacture a range of civil products in order to sustain itself in peacetime, very much demonstrating the characteristics of an armaments rather than an instrument manufacturer. Faced with the evaporation of such business, the firm's management turned to a two-pronged strategy for survival which embraced diversification and state subsidy.

The difficulties met by war industries in adapting to peace through diversification have been ascribed to their frequent lack of the relevant managerial and technical skills needed to move into unfamiliar markets. That this condition could be applied to Barr & Stroud might seem demonstrated by the lack of success in attempts to diversify into motorcycle engines and cinema projectors, but in fact the firm's philosophy in these efforts was quite different to other, much larger, arms businesses who sought to replace one activity with another. To Barr & Stroud, the issue was what could be made using existing capacity, which might be sold commercially to provide a bridge until enough government business came in to keep the firm employed in optical munitions manufacture. As early as 1919, the company's board was convinced that the future lay in doing what the business had always done, despite the problems confronting it. That judgement was based on instinct rather than a detailed study of options, but its soundness was shown by the subsequent events. The company succeeded in getting the Admiralty to believe that the symbiotic relationship that had evolved between them over the previous thirty years was balanced much in favour of the Royal Navy, and that the company had to be kept alive for the benefit of the service rather than the firm. Survival was not always a matter of making a demonstrably superior product. Although the Admiralty, or rather the Treasury, balked at an outright subsidy, the willingness of the Admiralty to guarantee a substantial level of profit on a reduced volume of business enabled Barr & Stroud to retain enough of their skilled workforce to keep going in the munitions business, and to be able to abandon its relatively un-remunerative civil ventures. By 1923, irrespective of the reduced levels of business, Barr & Stroud was tacitly recognized as an essential component in the Admiralty's establishment. It also remained, as in 1914, the world's only manufacturer devoted wholly to the production of optical munitions.

For the British optical munitions industry in the early 1920s, success became synonymous with survival. In 1888 there had been no industry, it had emerged after then in the wake of evolving armaments technologies and its products were only taken seriously by the British Army and the Royal Navy in the first years of the twentieth century. By 1914, still led by the evolution of military science, optical munitions were an integral part of strategic weapons systems and the importance of their makers to the state increasingly apparent. The First World War provided a totally unprecedented scale of demand and created a vastly expanded and vital industry that suffered an inevitable implosion with the armistice of 1918, threatening to leave the state without any means of producing the now-essential instrumentation of warfare. Such a simple summary implies a deterministic nature to the shaping of the industry, but as the preceding narrative has shown, much of that shaping was done under social and cultural influences that were no less important than the technology that framed them. Under those influences the optical munitions industry equipped not only Britain's armed forces but also most of the world's navies and not a few of its armies before 1914. It provided profits for its members, adapted to the needs of the greatest war yet experienced, and emerged into the 1920s in a drastically truncated but still capable form. That it was effectively reduced to only one company was not so much evidence of failure but an unavoidable adaptation to vastly changed circumstances. Faced with a major shift of international political attitudes towards armaments that looked likely to eliminate any large future demand for optical munitions, Barr & Stroud not only survived the transition from war to peace, but through adhering to a policy of austere specialization continued until the re-armament programmes of the 1930s simultaneously resurrected the demand for major items of optical munitions and stimulated the development of the electronic range and targeting systems that would eventually make them obsolete. 1923 marked not the end of the British optical munitions industry, but the start of a period of hibernation.

A Postscript

Having just announced what may seem very much like the advance notice for the eventual demise of optical munitions, it would nevertheless be wrong to give the impression that either optical munitions or their industry are now firmly and wholly in the past. Far from it. Armies still depend on prismatic binoculars as much as they did in the First World War, and more than ever before they use sighting telescopes to aim small arms and artillery. Whilst it is true that electronics in the form of radar displaced the optical rangefinder during and after the Second World War, range measuring on the world's battlefields is now once again being done with another type of optical device, the laser. Optical munitions and the industry that makes them are still very much in business.

NOTES

Introduction

1. Great Britain, Ministry of Munitions, *History of the Ministry of Munitions* (hereafter *OH*), 12 vols (London: HMSO, 1922).
2. R. MacLeod and K. MacLeod, 'Government and the Optical Industry in Britain 1914–1918', in J. M. Winter (ed.), *War and Economic Development* (Cambridge: Cambridge University Press, 1977), pp. 165–203, on p. 165.
3. Ibid., p. 191.
4. M. E. Williams, *The Precision Makers: A History of the Instruments Industry in Britain and France, 1870–1939* (London: Routledge, 1994). In particular, see chapters 2, 3 and 4.
5. Ibid., p. 1.
6. Ibid., p. 34.
7. A. McConnell, *Instrument Makers to the World: A History of Cooke, Troughton & Simms* (York: William Sessions Ltd, 1992).
8. Ibid., p. 64 and pp. 72–8.
9. J. T. Sumida. *In Defence of Naval Supremacy; Finance, Technology and British Naval Policy 1889–1914* (London, Routledge, 1993). See chapters 3, 5 and 6 for details on fire-control systems as a self-contained technological entity.
10. Ibid., pp. 72–6.
11. N. Friedman, *U.S. Submarines through 1945: An Illustrated Design History* (Annapolis, MD: Naval Institute Press, 1995), p. 267.
12. The quotation is from *Arms and Explosives*, October 1913, cited by C. Trebilcock, *The Vickers Brothers: Armaments and Enterprise 1854–1914* (London: Europa Publishers, 1977), p. 3.
13. Those interested in the make-up of the photographic lens industry in this period may care to entertain themselves by ploughing through *The Lens Collector's Vademecum*, available at the time of writing from http://antiquecameras.net/lensvademecum.html [accessed 28 January 2013].

1 The Emergence of the Industry, 1888–99

1. For descriptions of observation instruments, see W. Reid, 'Binoculars in the Army, Part 1: 1856–1903', *Army Museum* (1981), pp. 10–23 and R. J. Cheetham, *Old Telescopes* (Southport, Lancashire: Samedie, 1997), and for descriptions of early rangefinding devices, see Great Britain, Army, *Regulations for Musketry Instruction, 1896* (London: HMSO, 1896).

2. M. Moss and I. Russell, *Range and Vision: The First Hundred Years of Barr & Stroud* (Edinburgh: Mainstream Publishing, 1988), p. 13 documents the publication of the notice.

3. See W. F. Stanley, *Surveying and Levelling Instruments* (London: Spon, 1901), chs 1, 2 and 6 for survey techniques and instruments relevant to rangefinding.

4. Great Britain, Army, *Regulations for Musketry Instruction*, 1887 and 1896 edns (London: HMSO, 1887 and 1896). See tables E and F (1887) and fig. 419 (1896) for comparisons.

5. Great Britain, Army, School of Musketry, *Annual Report, 1893* (London: HMSO, 1893), p. 18.

6. The conditions are from the advertisement itself in University of Glasgow Records (hereafter UGD), UGD 295/16/1/4.

7. Extracted from R. G. W. Anderson, J. Burnett and B. Gee, *Handlist of Scientific Instrument Makers' Trade Catalogues 1600–1914* (Edinburgh: National Museums of Scotland, 1990) and the unclassified collection of makers' catalogues held by the National Media Museum, Bradford, West Yorkshire.

8. Moss and Russell, *Range and Vision*, pp. 22–3.

9. Ibid., ch. 1.

10. W. Stroud, *Early Reminiscences of the Barr and Stroud Rangefinders* (Glasgow: privately printed *c.* 1936), p. 6.

11. Captain Nolan, 'The Range-Finder', *Proceedings of the Royal Artillery Institution*, 2 (1874), pp. 161–207.

12. Moss and Russell, *Range and Vision*, pp. 18–21; L. C. Martin, *Optical Measuring Instruments: Their Construction Theory and Use* (London: Blackie, 1924), pp. 104–7.

13. Moss and Russell, *Range and Vision*, p. 21.

14. UGD 295/16/1/4, Barr to War Office, 13 June 1888.

15. UGD 295/16/1/4, Barr's personal papers, War Office to Barr, 16 June 1888, extract from War Office *Memorandum for Inventors*.

16. British Patent 9520/1888 records the application date as 30 June 1888.

17. UGD 295/16/1/1, Barr to Ripon, 1 January 1889 and UGD 295/16/1/17, Barr to Ordnance Committee, 18 March 1889.

18. UGD 295/16/1/17, Barr to Ordnance Committee, 18 March 1889 and A. Barr and W. Stroud, *Memorandum to the Ordnance Committee, Royal Arsenal Woolwich,* March 18 1889. UGD 295 'Un-classified papers'. Hereafter '*Memorandum*'.

19. UGD 295/16/1/4, Barr to Ordnance Committee acknowledging their decision, 21 January 1889.

20. Barr and Stroud, *Memorandum*, compares the two types in detail, and UGD 295/16/1/4, Barr's personal papers, Barr to Ordnance Committee acknowledging their decision, 3 June 1889 and Ordnance Committee to Barr, 19 June 1889.

21. For prism designs see G. Smith and D. A. Atchison, *The Eye and Visual Optical Instruments* (Cambridge: Cambridge University Press, 1997), ch. 8. For other information on optical systems, see Martin, *Optical Measuring Instruments* and D. F. Horne, *Optical Instruments and their Applications* (Bristol: Adam Hilger Ltd, 1980).

22. Moss and Russell, *Range and Vision*, p. 23, and Stroud, *Early Reminiscences*, p. 8.

23. Stroud, *Early Reminiscences*, p. 8.

24. Moss and Russell, *Range and Vision*, p. 22.

25. For Hilger's history see Horne, *Optical Instruments and their Applications*, p. 34. For connections with Campbell see Science Museum Library, Adam Hilger collection (hereafter HILG), HILG 3/1, History, 'Notes from Mr Johnson in connection with his history of Adam Hilger', 6 November 1952 and HILG 3/1, 'Mr Twyman's Lecture, August 1944'.

26. UGD 295/16/1/3, Barr, personal correspondence, William Thomson to Barr, 8 March 1889, commenting on the design.
27. HILG 3/1, Sales Manager, Hilger & Watts (successor company to Adam Hilger Ltd) to Barr & Stroud Ltd, 25 January 1952.
28. Moss and Russell, *Range and Vision*, p. 23.
29. Nolan, 'The Range-Finder', p. 162.
30. E. G. Edwards, 'Field Range-Finding', *Proceedings of the Royal Artillery Institution*, 11 (1883), pp. 202–14.
31. The National Archives, Kew, London (hereafter TNA), WO 32/8902.
32. McConnell, *Instrument Makers to the World*, p. 64.
33. Great Britain, Army, *Handbook of the Mekometer* (London: HMSO, 1911), pp. 9–13 specifies the instrument's operation and provides the subsequent description.
34. Great Britain, Army, School of Musketry, *Annual Report 1893*, p. 18.
35. M. J. Bastable, *Arms and the State: Sir William Armstrong and the Remaking of British Naval Power, 1854–1914* (Aldershot: Ashgate Publishing, 2004), pp. 68–9.
36. McConnell, *Instrument Makers to the World*, p. 64. The award was worth approximately £2.8 million in 2011 values. See online at http://www.measuringworth.com/ukcompare/relativevalue.php [accessed 10 January 2013].
37. McConnell, *Instrument Makers to the World*, p. 61.
38. Ibid., p. 65.
39. HSBC Bank Group Archives, *Abstract of Entries in the Yorkshire Banking Company Board Minutes relating to T. Cooke & Sons*, 1882–1900, references X19 to X24, provides the source for the financial details in this section.
40. University of York, Borthwick Institute of Historical Research, Vickers Instruments Archive, Company Records of Cooke, Troughton & Sims (hereafter VIA), AJB 070/1.3/ box 1, drawings index 1882–1921, drawings number 369 and 370.
41. For instruments in service, see Great Britain, Army, School of Musketry, *Annual Report, 1893*, p. 18 and for obsolescence of earlier types see Great Britain, Army, *List of Changes in War Material* (London: HMSO, 1891), entry dated 24 August 1891.
42. The scale of issue and cost are supplied by TNA, Records of the Treasury TI/11223, Proceedings of the Ordnance Council, 12 June 1908, p. 4.
43. This figure assumes a gross margin of 33 per cent of the selling price, a figure typical of commercial sales at the time.
44. C. E. Callwell and J. F. Headlam, *History of the Royal Artillery*, 3 vols (London: Royal Artillery Institution, 1937) vol. 1, pp. 204–5.
45. H. T. Seeger, *Feldstecher: Fernglaser Im Wandel Der Zeit* (Borken: Bresser-Optik, 1989), ch. 1.
46. For background to the 1889 Act, see Trebilcock, *The Vickers Brothers*, pp. 52–5.
47. For details of the new guns, see E. W. Lloyd and A. G. Hadcock, *Artillery: Its Progress and Present Position* (Portsmouth: J. Griffin & Co., 1893), ch. 7.
48. Hampshire County Record Office, Priddy's Hard Material, collection reference 109/M/91 (hereafter HCRO) HCRO/PQ2, Great Britain, Admiralty, Gunnery Department. *Monthly Record of Principal Questions Dealt with by the Director of Naval Ordnance*, July–December 1889 provides the source material for this section unless otherwise indicated.
49. Moss and Russell, *Range and Vision*, p. 25.
50. HCRO/PQ6, Recommendation of Director of Naval Ordnance, 9 February 1892.

51. A. Pollen, *The Great Gunnery Scandal* (London: William Collins & Co. Ltd, 1980), pp. 66, 260.
52. See ibid., ch. 2 and Sumida, *In Defence of Naval Supremacy*, ch. 5 for attitudes towards shipboard instrumentation.
53. For details on the uses of telescopes, see PRO ADM 116/407, 'Long distance telescopes: trials and issue to H. M. Ships', 1893 to 1896.
54. UGD 295/16/1/5, correspondence, Admiralty to Professor Barr, 10 June 1892.
55. Moss and Russell, *Range and Vision*, p. 26. The sum was equal to approximately £8.3 million in 2011 values. See online at http://www.measuringworth.com/ukcompare/ relativevalue.php for figures using the GDP deflator [accessed 10 January 2013].
56. UGD 295/4/11, Letter book, Barr to Director of Naval Contracts acknowledging receipt of Admiralty acceptance of tender for contract CP NS4886–927383/1620.
57. Moss and Russell, *Range and Vision*, pp. 25–31 provides source material for this section unless otherwise indicated.
58. UGD 295/22/1/8, includes British Patents 11025/1889, 4185/1890, 12448/1890, 12736/1890 and 3172/1891.
59. UGD 295/4/11, Letter book 1893, contains a series of eighteen letters from Barr to Chadburn's ordering an assortment of lenses.
60. Anon., *The Century's Progress: Yorkshire Industry and Commerce 1893* (London: The London Printing and Engraving Co., 1893, reprint, Brenton Publishing, 1971), p. 141.
61. UGD 295/16/1/5, correspondence, Strang papers, Stroud to Hilger on methods of making pentagonals for the rangefinder, 6 October 1890 and UGD 295/4/11, Letter book, Barr to Hilger complaining of incorrect angles of pentagonals.
62. Moss and Russell, *Range and Vision*, p. 29.
63. Stroud, *Early Reminiscences*, marginal note in Stroud's own handwriting on p. 12 of the copy in UGD 295/26/1/55.
64. UGD 295/4/11, Letter book, Barr to Armstrong Mitchell quoting for supply of thirteen rangefinders at £700 less 12.5 per cent commission.
65. UGD 295/4/11, Barr to Armstrong Mitchell, 9 May 1893, and Barr to Armstrong Mitchell, 3 July 1893.
66. Lloyd and Hadcock, *Artillery*, p. 9 for the rangefinder and p. 10 for spotting the fall of shot as an aid to ranging.
67. UGD 295/4/11, Barr to Armstrong Mitchell, 3 July 1893.
68. Stroud, *Early Reminiscences*, p. 10.
69. This figure is taken from J. M. Strang's research material for the unpublished history of Barr & Stroud cited by Moss and Russell. Both the material and typescript are in UGD 295, unclassified material, Strang papers.
70. Extracted from UGD F2/16, Ledger 2, pp. 80–7.
71. UGD 295/4/11, Letter book, Barr to Armstrong Mitchell, 7 July 1893 and 9 August 1893 detail the terms; 28 September 1893 confirms signature of agreement.
72. No reason for this has been found in the company's records.
73. For the influence that Royal Navy had on foreign powers, see Trebilcock, *The Vickers Brothers*, chs 3 and 4.
74. UGD 295/4/12, Letter book, Jackson to Admiralty, 26 May 1894.
75. UGD 295/4/13, Letter book, Jackson to Armstrong Mitchell.
76. UGD 295/16/1/13, Foreign letters, German Embassy, London, to Barr & Stroud, provides source material for the following quotations in this paragraph, 2 April 1894, 20 April 1894, 23 June 1894 and 12 July 1894.

77. UGD 295/4/12, Letter book, Barr to Captain Tülick, 23 April 1894 provides the source material for the rest of this paragraph.
78. Stroud, *Early Reminiscences*, p. 14.
79. Figures extracted from UGD 295/26/1/93, Dr W. Strang's personal papers, and UGD 295/26/1/27 and 295/26/1/28, 'historical notes' prepared for Dr Strang's proposed history of the firm.
80. UGD 295/4/12, Letter book, Barr to Captain Hall on HMS *Resolution*, compares differences between individual rangefinders delivered, 27 August 1894.
81. Moss and Russell, *Range and Vision*, p. 33.
82. UGD 295/4/13, Letter book, Barr & Stroud to Director of Naval Contracts, tender dated 22 April 1895.
83. Moss and Russell, *Range and Vision*, p. 31.
84. UGD 295/16/1/5, Admiralty to Barr, 1 November 1893.
85. UGD 295/4/12, Letter book, Barr to Admiralty, 25 April 1894; UGD 295/4/13, Letter book, Barr to Armstrong Mitchell, 7 May 1895.
86. Moss and Russell, *Range and Vision*, p. 24.
87. Ibid., pp. 31–3.
88. UGD 295/4/13, Letter book, Barr to Armstrong Mitchell, 24 April 1895, 29 April 1895 and 5 May 1895.
89. UGD 295/4/14, Letter book, Barr to Armstrong Mitchell, 9 April 1895, 6 November 1895 and 20 January 1896.
90. Moss and Russell, *Range and Vision*, p. 29.
91. UGD 295/4/14, Letter book, Barr to Capt. W. S. Cowles, US Navy, US Legation London, announcing his proposed itinerary; Barr to Armstrong's from New York announcing he was there on university business, 24 April 1896; Barr to Jackson, 24 June 1896 and Barr to Colonel Ludlum, 13 April 1896.
92. UGD 295/16/1/9, Archibald Barr, personal correspondence, Barr to Stroud, 26 January 1897 provides the source material for the rest of this section, unless otherwise indicated.
93. Moss and Russell, *Range and Vision*, pp. 34, 35.
94. UGD 295/16/1/2, Typescript note initialled 'JWF' on envelope labelled 'Letters from Dr Barr of period 1897–98' in Barr's personal papers.
95. UGD 295/4/15, Letter book, Barr to Armstrong's, 26 April 1897.
96. Moss and Russell, *Range and Vision*, p. 34.
97. Ibid.
98. Figures extracted from UGD 295/26/1/93, Dr Strang's personal papers, and UGD 295/26/1/27 and 295/26/1/28, 'historical notes' prepared for Dr Strang's proposed history of the firm.

2 The Growth in Importance from the Boer War to 1906

1. For details of the artillery weapons used, see H. C. B. Rodgers, *Artillery through the Ages* (London: Seeley, Service, 1971), ch. 10.
2. TNA WO 395/1, *Quantities of Scientific Instruments purchased 1 April 1899 to 31 March 1902*, p. 68.
3. TNA WO 108/278, *Extracts from Reports by Officers Commanding Units in South Africa during 1899–1901*. Figures extracted from section dealing with signalling equipment, telescopes and binoculars where 603 officers had been asked for responses.
4. Callwell and Headlam, *History of the Royal Artillery*, vol. 2, ch. 3.

5. Ibid., vol. 2, p. 46 supplies content about South Africa. For the method of using the Mekometer, see Great Britain, Army, *Regulations for Musketry Instruction, 1896*.
6. Stroud, *Early Reminiscences*, p. 9, refers to high casualties; G. Forbes, *Experiences in South Africa with a New Infantry Range-finder* (London: J. J. Keliher & Co. Ltd, 1902), p. 4 refers to the Mekometer's lack of use.
7. C. Trebilcock, 'War and the Failure of Industrial Mobilisation', in J. M. Winter (ed.), *War and Economic Development: Essays in Memory of David Joslin* (Cambridge: Cambridge University Press, 1975), p. 139–64.
8. MacLeod and MacLeod, 'Government and the Optical Industry in Britain', p. 168.
9. Ibid., pp. 168–9.
10. C. Trebilcock, 'War and the Failure of Industrial Mobilisation', pp. 141–2.
11. Ibid., pp. 149–51.
12. These procedures remained unchanged until after the First World War began in 1914: for details see *OH*, vol. 1, part 1, pp. 53–8.
13. UGD 295/4/23, Letter book, H. D. Jackson to War Office stating that Barr & Stroud could produce those types if asked, 30 July 1901. This provides the source material for the rest of this section, unless otherwise indicated.
14. Ottway & Co. made sighting telescopes, Broadhurst & Clarkson made observation telescopes, The Ross Optical Co. and W. Watson & Sons both made observation and sighting telescopes.
15. UGD 295/4/24, Letter book, Jackson to Director of Army Contracts, 12 February 1902.
16. UGD 295/19/2/1, Customer Order files 1901–3, order numbers CO 193, 215, 235, 245, 291, 294 and 348.
17. Nowhere in the firm's records is there any report of the War Office making detailed enquiries about its capacity, or of any official visit to the factory.
18. Details extracted from PRO WO 395/1, Annual Reports of the Director of Army Contracts, Financial Years ending 31 March 1899 to 31 March 1902.
19. TNA WO 395/1, Annual Report of the Director of Army Contracts 1901–2, p. 68.
20. TNA WO 395/1, Annual Report 1902–3, p. 88.
21. TNA WO 108/278, *Extracts from Reports by Officers Commanding Units in South Africa during 1899–1901*. Figures extracted from section dealing with signalling equipment, telescopes and binoculars provides the data regarding local purchases.
22. Great Britain, School of Musketry, *Annual Report 1893*, p. 89.
23. A. Gleichen, *The Theory of Modern Optical Instruments: A Reference Book for Physicists, Manufacturers of Optical Instruments and for Officers in the Army and Navy*, trans. H. Emsley and W. Swain (London: HMSO, 1918), p. 196.
24. Moss and Russell, *Range and Vision*, p. 42, and Forbes, *Experiences in South Africa with a new Infantry Range-finder*.
25. Callwell and Headlam, vol. 2, pp. 107, 108.
26. UGD 295/4/744, Letter book 1897–1911, Barr & Stroud to War Office asking for details of WO requirements, 29 August 1899.
27. Callwell and Headlam, vol. 2, p. 108 provides the source material for the rest of this paragraph.
28. TNA TI/11223, Proceedings of the Ordnance Council, 12 June 1908, Question of Reward to Captain A. H. Marindin, p. 3.
29. TNA TI/11223 (1908), p. 8.
30. UGD 295/16/1/10, letter to H. D. Jackson, 26 May 1902.

31. UGD 296/16/1/10, Stroud to Jackson, 26 July 1901.
32. For an explanation of this principle, see F. Auerbach, *The Zeiss Works and the Carl Zeiss Stiftung in Jena*, trans. F. Cheshire and S. Paul, 2nd edn (London: Marshall Brookes & Chalkely, 1904), p. 66, 67 and Smith and Atchison, *The Eye and Visual Optical Instruments*, pp. 450, 451.
33. UGD 296/16/1/10, Stroud to Jackson, 27 November 1901 and 29 November 1901.
34. UGD 296/16/1/10, Jackson to Stroud, 29 November 1901.
35. VIA AJB, 210.2.5, lecture to the Society of Arts, 18 December 1901, paper read to the Royal Society, 20 March 1902 and lecture to the Royal United Service Institution, 13 May 1902. Professor G. Forbes, 'Experiences in South Africa with a new Infantry Range-finder' in *Journal of the Royal United Service Institution*, 13 May 1902, provides the source for the rest of this section, unless otherwise indicated.
36. UGD 296/16/1/10, Stroud to Jackson, 6 July 1902.
37. Science Museum London, Library, Adam HILG 3/1, Typescript of Mr Twyman's Lecture, August 1944, p. 15.
38. HILG 3/1, p. 24.
39. HILG 3/1, p. 15.
40. UGD 295/4/21, Letter book, H. D. Jackson to Hilger, a series of letters between 3 October 1900 and 3 December 1900 describes how relations fluctuated.
41. TNA TI/11223 (1908), p. 8
42. TNA TI/11223 (1908), p. 13, List of payments received by Adam Hilger Ltd and TNA TI/11223 (1908), p. 3.
43. TNA TI/11223 (1908), p. 3.
44. UGD 295/4/744, Letter book, Barr's reply to Forbes' undated proposal, 2 October 1902 provides source material for the rest of this paragraph.
45. See Patents, Designs and Trade Marks Act 1883, Section 27(2) which refers to terms between inventors and the Crown. The patents for the prismatic binocular component of Forbes's design were actually held by Zeiss.
46. TNA TI/11223 (1908), p. 3.
47. For Barr & Stroud's opinion, see GUA UGD 295, Unclassified material, Russell Research Notes: 9 January 1903, H. D. Jackson to Archibald Barr. Russell cites a Letter book 'BS4/21'. Russell's notes were made before the University acquired the Barr & Stroud records, and some of the material quoted is not now in the University Archives. For Twyman's interpretation, see HILG 3/1, p. 15.
48. HILG 3/1, p. 15.
49. TNA TI/11223 (1908), p. 12 and TNA TI/11223, Proceedings of the Ordnance Council, 8 June 1909, pp. 5, 10.
50. UGD 295, Strang papers.
51. D. K. Brown, *The Grand Fleet: Warship Design and Development 1906–1922* (London: Chatham Publishing, 1999), p. 26.
52. H. Garbett, *Naval Gunnery* (London: George Bell, 1897), pp. 201–3.
53. Gun Sighting Telescope type AP 360 had been introduced in 1887, and type AP 700 in 1891: PQ 109/M/91/PQ11 details the instructions not to use the sights.
54. Sumida, *In Defence of Naval Supremacy*, p. 46.
55. TNA ADM 116/602 Naval Armaments and Equipment; experiences gained on active service in South Africa by the Naval Brigade, has comments on the quality of naval telescopic sights.
56. PQ 109/M/91/PQ16, 16 February 1905.

57. PQ 109/M/91/PQ16, 2. June 1905.

58. This figure has been extrapolated from the scales of issue in the schedule, and from armament details in *Jane's Fighting Ships*, 1905–1906.

59. PQ 109/M/91/PQ16, 16 February 1905. They are not mentioned at all in the Royal Navy's *Manual of Gunnery for H. M. Fleet*, 1:1 (London: HMSO, 1907).

60. Great Britain, Admiralty, Gunnery Department, *Manual of Gunnery for H. M. Fleet* (London: HMSO, 1907).

61. *Rate Book for Naval Stores: Authorised List and Price List of Naval Stores* (HMSO, annually from 1870).

62. Figure arrived from data supplied by http://www.measuringworth.com/index.php using the GDP deflator [accessed 3 January 2013].

63. Extracted from UGD 295/19/1/2, Customer Orders 1900–1910.

64. UGD 295/5/744, Letter book, Barr & Stroud to Admiralty, 30 May 1898, cites Admiralty letter CP/4919/8720, 27 April 1898.

65. UGS 295/5/744, Letter book, Barr & Stroud to Admiralty, 30 May 1898.

66. Trebilcock, *The Vickers Brothers*, p. 3.

67. Patents, Designs, and Trade Marks Act, 1883, Section 27(1).

68. The provisions of British Patent 9520/1888 were also in force in France, Germany and the US.

69. Moss and Russell, *Range and Vision*, p. 37.

70. Ibid., p. 42 provides the source for figures and quotations in this section.

71. UGD 295 16/1/10, personal correspondence of William Stroud. Harold Jackson had been made a junior partner in the business by then.

72. UGD 295/16/1/10, personal correspondence, Stroud to Barr, 16 March 1899.

73. UGD 295/16/1/10, personal correspondence, Stroud to Jackson, 7 July 1899.

74. Moss and Russell, *Range and Vision*, p. 43.

75. UGD 295/4/12 Letter book, Archibald Barr to Secretary, Admiralty, 25 April 1894.

76. UGD 295/4/23 Letter book, Barr to Stroud, 7 September 1901.

77. Extracted from UGD 295/19/2/1 Customer Orders 1901 to 1910.

78. UGD 295/19/2/1 Customer Order files 1901–10.

79. Moss and Russell, *Range and Vision*, p. 43.

80. P. Padfield, *Aim Straight: A Biography of Sir Percy Scott* (London: Hodder & Stoughton, 1966), p. 135.

81. For the relative state of gunnery in the two navies, see H. W. Wilson, *Battleships in Action*, 2 vols (London: Conway Maritime Press, 1995), vol. 1, ch. 11; Russian order details extracted from UGD 295/19/2/1, Customer Order files 1901–10.

82. Moss and Russell, *Range and Vision*, p. 45 provides source material for the rest of this section, unless otherwise indicated.

83. UGD 295, Unclassified material, Strang manuscript, p. 61.

84. UGD 295/4 /739, Letter book, Barr to Stroud, 16 November 1904.

85. UGD 295/4/739, Letter book, Barr to Stroud, 14 December 1904.

86. In March 1901 he refused to travel up to Glasgow, citing the presence of smallpox in the city as too great a risk. UGD 295/16/1/10, Stroud's personal correspondence.

87. See T. N. Clark, A. D. Morrison-Low, and A. D. C. Simpson, *Brass & Glass: Scientific Instrument Making Workshops in Scotland* (Edinburgh: National Museums of Scotland, 1989) for an account of this business and the connections with Lord Kelvin.

88. UGD 295, Unclassified material, Russell research notes: Private Ledger No. 1 (1888–1902) is mentioned as showing payments to Chadburn Brothers from 1889 onwards.

89. For example, UGD 295/4/22, Letter book, Barr & Stroud to Chadburn Bros, 25 February 1901 complains that 'in almost every instance … we get a wrong lens'.

90. Clark et al. (1990) gives no listing of optical manufacturers in or near Glasgow.

91. For a description of prisms and their working in the rangefinder, see Martin, *Optical Measuring Instruments*, p. 113.

92. UGD295/4/11 Letter book, Barr to Hilger, 23 March1893 and 4 April1893.

93. UGD 295/16/1/9, Letter book, Barr to Stroud, 26 January 1897.

94. UGD 295, Unclassified Material, Russell research notes, H. D. Jackson to Hilger, 10 September 1898.

95. UGD 295/4/21, Letter book, H. D. Jackson to Hilger, a series of letters between 3 October1900 and 3 December 1900.

96. UGD 295/16/1/9, Letter book, Barr to Hilger, 13 December 1900.

97. UGD 295/4/17, Letter book, Barr & Stroud to Ross Optical Co., requesting them to tender for telescope objectives 12 December 1898.

98. For example, see UGD 295/4/22, Letter book, Jackson to Goerz, 28 March 1901, requesting quotations for 'fine quality prisms'.

99. UGD 295/4/17, Letter book, ten objective lenses ordered on 31 December 1898. They were returned as faulty on 6 March1899 and were the subject of a subsequent dispute.

100. UGD 295/16/1/10, personal correspondence from William Stroud to H. D. Jackson, 7 July 1899.

101. UGD 295, Unclassified material, Russell research notes, Barr's Private Letter Book No. 1, Barr to Mr A. Hilger [*sic*], 19 January 1904.

102. N. A. Lambert, *Sir John Fisher's Naval Revolution* (Columbia, SC: University of South Carolina Press, 1999), p. 38.

103. M. F. Suetter, *The Evolution of the Submarine Boat, Mine, and Torpedo* (Portsmouth: Griffin, 1908) describes these in some detail in the text.

104. Cited in I. S. Glass, *Victorian Telescope Makers: The Lives and Letters of Thomas and Howard Grubb* (Bristol: Institute of Physics Publishing, 1998), p. 206.

105. Glass, *Victorian Telescope Makers* describes the other accomplishments of the firm.

106. Ibid., p. 208.

107. Friedman, *U.S. Submarines through 1945*, p. 270.

3 Expansion and Consolidation, 1907–14

1. TNA War Office records WO 395/2, Annual Reports of the Director of Army Contracts, financial years 1903–1904, 1904–1905, 1905–1906, 1906–1907; data extracted from Contracts for Scientific Instruments.

2. MacLeod and MacLeod, 'Government and the Optical Industry in Britain', p. 170.

3. Extracted from TNA WO 395/2 and WO 395/3, Annual Reports.

4. TNA TI/11223, Proceedings of the Ordnance Council, 12 June 1908, *Question of Award to Captain A. H. Marindin, The Black Watch for One-man Range-finder for Infantry* (hereafter TI/11223 1908), p. 5.

5. TNA TI/11223 (1908), p. 4. This provides the source material for the rest of this section, unless otherwise indicated.

6. TNA TI/11223, Proceedings of the Ordnance Council, 8 June 1909, *Question of Award to Captain A. H. Marindin, The Black Watch for One-man Range-finder for Infantry* (hereafter TI/11223 1909), p. 6.

7. University of Glasgow Archives, University of Glasgow Archives, Barr & Stroud, City of Glasgow, Optical Instrument Makers, collection reference UGD 295 (hereafter UGD 295) 295/4/744, Letter book, H. D. Jackson to Adam Hilger Ltd, 9 August 1907 and 20 November 1907.
8. TNA TI/11223 (1908), p. 4, 10.
9. TNA WO 395/3, Contracts Department, Annual Report 1907–1908.
10. Extracted from UGD 295/19/2/1, 295/19/2/2 and 295/19/2/3, Customer Order files.
11. MacLeod and MacLeod, 'Government and the Optical Industry in Britain', p. 170.
12. Callwell and Headlam, *The History of the Royal Artillery*, vol. 2, chs 3 and 10.
13. Ibid., vol. 2, pp. 95–101 describe the problems of indirect firing and aiming and provide the source material for the rest of this paragraph.
14. See B. K. Johnson, 'The No. 7 Dial Sight, Mk. 2'. *Transactions of the Optical Society*, 21:5 (1920), pp. 176–86.
15. A. Hagen, *Deutsche Direktinvestionen in Grossbritannien, 1871–1918* (Stuttgart: Steiner Verlag, 1997), pp. 174–5 provides information about Goerz, and A. Hagen 'Export versus Direct Investment in the German Optical Industry', *Business History*, 4 (October 1996), pp. 1–20, on p. 5 details the effects of new legislation on German companies trading in Great Britain.
16. Extracted from TNA WO 395/2 and 395/3, Annual Reports of Director of Army Contracts.
17. For numbers of guns ordered, see I. V. Hogg and L. F. Thurston, *British Artillery Weapons and Ammunition 1914–1918* (Shepperton: Ian Allen Ltd, 1972), pp. 58, 80, 102 and 116, and for introduction into service see Callwell and Headlam, *The History of the Royal Artillery*, vol. 2, p. 101.
18. Extracted from TNA WO 395/3 Director of Army Contracts, *Annual Reports*, except for the Barr & Stroud order 1909–10, which comes from UGD 295/19/8/1, Customer Order records, Works Order CO 1115, 2 January 1910.
19. McConnell, *Instrument Makers to the World*, p. 65 describes the connections with Vickers.
20. TNA TI/11223 (1908), p. 4.
21. Trebilcock, *The Vickers Brothers*, p. 12 and then p. 11.
22. Ibid., p. 14.
23. MacLeod and MacLeod, 'Government and the Optical Industry in Britain', p. 170.
24. This proportion is obtained from TNA TI/11223.
25. These and the following figures for War Office businesses are extracted from TNA WO 395/3 Annual Reports of the Director of Army Contracts.
26. Hagen, 'Export versus Direct Investment in the German Optical Industry', p. 6.
27. Some of the orders are listed in the Director of Contracts' Annual Reports as being shared with other companies. Assuming an equal share of contract values produces this figure.
28. D. G. Hermann, *The Arming of Europe and the Making of the First World War* (Princeton, NJ: Princeton University Press, 1996), p. 234.
29. Sumida, *In Defence of Naval Supremacy*, pp. 185–96.
30. UGD 295, Unclassified material, Russell research notes refer to Barr & Stroud Letter book BS4/25: J. B. Henderson, Naval College, Greenwich, to Harold Jackson, 5 October 1907, describing the problems for rangefinding caused by vibration in destroyers.
31. Great Britain, Admiralty, *Manual of Gunnery for H. M. Fleet, 1915* (London, HMSO, 1915), vol. 1, ch,. 12, p. 291–304 describe and illustrate various patterns.
32. Great Britain, Admiralty, *Manual of Gunnery for His Majesty's Fleet, 1917* (London: HMSO, 1917), ch. 12.

33. See Sumida, *In Defence of Naval Supremacy*, ch. 5 for background material.

34. Great Britain, Admiralty, Gunnery Department, *Manual of Gunnery for H. M. Fleet, 1917*, p. 423.

35. Sumida, *In Defence of Naval Supremacy*, ch. 3.

36. UGD 295 Russell Research Notes box 2: letter from J. B. Henderson, Admiralty Research Laboratory, to William Stroud, 5 October 1907.

37. UGD 295 Russell Research Notes box 2: Acknowledgement of order for a 4.57 m range-finder from Col. Petrov, Imperial Russian Navy.

38. British Patents 1436/1901 for an optical rifle sight, and 23038/1903, 12735/1904 and 12902/1905 for rangefinders.

39. VIA AJB 070 1.3, Drawing Office Index, drawings 1674/5, 1873, 1885/6, 1898 and 2175.

40. Sumida, *In Defence of Naval Supremacy*, p. 85.

41. British Patents 7322/1907, 13562/1907, 15200/1907, 20315/1908, 6082/1910, 7392/1910 and 9306/1911.

42. See British Patent 30090/1912 for details of the instrument, and McConnell *Instrument Makers to the World*, p. 74 for a summary of the evolution of Pollen's system.

43. McConnell, *Instrument Makers to the World*, ch. 7 describes the range of Cooke's engineering activities.

44. VIA AJB 030/1.1.1, T. Cooke & Sons Ltd, Directors' minute book entry, Annual General Meeting 1908.

45. McConnell, *Instrument Makers to the World*, p.65.

46. UGD 295/4/1, Letter book, H. D. Jackson to J. Heather, 24 December 1908.

47. UGD 295/4/3, Letter book, H. D. Jackson to J. Heather, 21 March 1911.

48. TNA ADM 116/3458, correspondence between Admiralty and the Treasury on the need for only British optical glass to be used in Royal Navy instruments: Admiralty Report 27 August 1915, noting correspondence 1910 to 1913.

49. Moss and Russell, *Range and Vision*, p. 38.

50. UGD 295, Unclassified material, Russell research notes, box 2, Henderson to Stroud, 5 October 1907.

51. UGD 295/4/109, Letter book, J. W. French to Adm. Mouton, Royal Dutch Navy, reviewing progress and attitudes, 29 July 1914.

52. UGD 295/4/80, Letter book, J. W. French to W. Stroud, 25 March 1912.

53. For examples of Pollen's problems, see Pollen, *The Great Gunnery Scandal*.

54. UGD 295/4/53, Letter book, Jackson to Director of Naval Construction, 26 September 1908 and 30 September 1908.

55. UGD 295/4/4, Letter book, Jackson to Director of Naval Construction, 30 November 1912.

56. UGD 295/4/4, Jackson to Director of Naval Construction, 7 December 1912.

57. UGD 295/4/88, Letter book, Jackson to Capt. A. H. Seibert, 20 July 1912, saying he believed Pollen was Argo's 'designer'. This letter is also the source material for the rest of this paragraph.

58. E. W. Taylor, 'The New Cooke–Pollen Rangefinder', *Journal of the United States Artillery*, 41:3 (May–June 1914), p. 813[en]31. This article describes and illustrates the instrument.

59. Auerbach, *The Zeiss Works and the Carl Zeiss Stiftung in Jena* describes the advances in optical design in this period and the benefits of the new glasses that were constantly being introduced.

60. VIA AJB 220/2.6: H. D. Taylor's 1895 paper on the adjusting and testing of telescope objectives made clear his espousal of the new types.

61. TNA ADM 116/3458, correspondence between Admiralty and Treasury on optical glass supplies, 1910–1913 provides the source material for the following section unless otherwise indicated.

62. J. F. Chance, *A History of the Firm of Chance Brothers & Co.* (London: Ballantyne & Co. Ltd, 1919), p. 182–4.

63. UGD 295/4/107, Letter book, Jackson to S. Vronski, quoting from an unidentified Royal Navy officer, 14 May 1914.

64. See W. McBride, *Technological Change and the U.S. Navy 1865–1945* (Baltimore, MD: Johns Hopkins University Press, 2000).

65. The figures up to 1912 have been extracted from UGD 295/26/1/27 'Correspondence and notes', and UGD 295/26/1/47, Table of Barr & Stroud sales, 1901–1912. The later figures come from UGD 295/11/1, Balance sheets for 1913.

66. Extracts from UGD 295/19/2/1, 295/19/2/2 and 295/19/2/3, Customer Order files 1907–14.

67. UGD 295/4/46, Letter book, provides the source material for the following section unless otherwise indicated.

68. The agreement is referred to in Jackson's letter of 20 November, but no details of it were given.

69. Moss and Russell, *Range and Vision*, p. 38.

70. UGD 295/4/41, Letter book, Jackson to Hilger, 7 June 1906. See also index entries in other letter books in the UGD 295/4 series for evidence of the volume of correspondence.

71. UGD 295, Unclassified material, Russell research notes, folder 'Private Letter Books, BS4/21', Stroud to Barr, 24 January 1907 ('BS4/21' is not listed among the re-classified material now in the UGD 295 collection.)

72. See chapter 1 and 2 of this volume.

73. UGD 295/4/744, Letter book, Barr to Twyman expressing condolences on Otto Hilger's death, and explaining the sense of loss he felt, 20 December 1902.

74. UGD 295/4/80, Letter book, Jackson to Chance Brothers, urging delivery of pentagonal blocks; Jackson to Twyman on wage rates for polishers, 3 April 1912; Jackson to P. de Braux 3 April 1912.

75. Science Museum Library, Hilger Collection, HILG 3/1, Typescript of 'Mr Twyman's lecture, August 1944'.

76. UGD 295/4/81, Letter book, Jackson to Dallmeyer, 23 April 1912, to Ross 23 April 1912, to TT&H 23 April 1912, to Chadburn 6 April 1912.

77. UGD 295/4/81, Jackson to P. de Braux, 3 April 1912.

78. UGD 295/4/86, Letter book, Jackson to J. W. French, 26 November 1912.

79. UGD 295/4/95, Letter book, Jackson to Watson, 12 May 1913, and Jackson to Moeller, 14 May 1913.

80. See also S. C. Sambrook, 'The British Armed Forces and their Acquisition of Optical Technology' in *Yearbook of European Administrative History* (Nomos Verlagsgesellschaft, Baden-Baden, 2008), pp 139–164.

81. Lambert, *Sir John Fisher's Naval Revolution*, p. 38.

82. Suetter, *The Evolution of the Submarine Boat, Mine, and Torpedo*, describes these in chapter 19.

83. Cited in Glass, *Victorian Telescope Makers*, p. 206.

84. A. N. Harrison, *The Development of HM Submarines* (London: Ministry of Defence, 1979), p. 22.9.
85. Ibid., pp. 22.4–22.11.
86. Little is yet known about this firm's activities in periscope manufacture. Their records in Glasgow University Archives (UGD 33/4) make it plain that periscopes were being made, but give little detail as to how the involvement arose. There was no relationship with Barr & Stroud.
87. Harrison, *The Development of HM Submarines*, p. 22.6.
88. See Williams, *The Precision Makers*, ch. 2 and MacLeod and MacLeod, 'Government and the Optical Industry in Britain'.
89. Trebilcock, *The Vickers Brothers*, pp. 1–7 provides these and the following characteristics.

4 The Impact of War, August 1914 to mid-1915

1. British Science Guild; *Report of the Technical Optics Committee respecting the Proposed Establishment of an Institute of Technical Optics*, June 1914, appendix C to the *Ninth Annual Report of the BSG* (June 1915), pp. 29–31. Cited in MacLeod and MacLeod; 'Government and the Optical Industry in Britain', p. 170.
2. *OH*, vol. 11, 'The Supply of Munitions', part 3 (London: HMSO, 1922), p.13, p. 9 and p. 18.
3. MacLeod and MacLeod, 'Government and the Optical Industry in Britain', p. 170.
4. Williams; *The Precision Makers*, p. 8.
5. Extracted from London Metropolitan Archives, LCC/MIN/2967 (Papers accompanying Reports 1911) *Report of the Education Committee of the London County Council: Proposed Establishment of a Technical Institute for Optics* 1 March 1911 (hereafter LCC Report 1911).
6. MacLeod and MacLeod, 'Government and the Optical Industry in Britain', p. 175.
7. For details of the organization of private sector small-arms production, see *OH*, vol. 11, part 4, 'Rifles', pp. 3–21.
8. Details extracted from TNA, Records of the War Office, WO 395/3, Director Army Contracts Annual Report 1914.
9. Extracted from TNA 395/3, Director of Army Contracts Annual Reports, 1912, 1913 and 1914.
10. Hagen; 'Export versus Direct Investment in the German Optical Industry', see also ch. 4 in this volume.
11. *OH*, vol.1, part 1, appendix 2, p. 145, and Hermann, *The Arming of Europe and the Making of the First World War*, p. 234.
12. *OH*, vol. 1, part 1, p. 8 provides the source material for the rest of this paragraph.
13. Ibid., pp. 9 and 10 provide the data for the rest of this section on the enlargement of the Army.
14. H. Strachan, *The First World War: Volume 1: To Arms* (Oxford: Oxford University Press, 2001), p. 1067.
15. MacLeod and MacLeod, 'Government and the Optical Industry in Britain', p. 171 and A. C. Williams, 'The Design and Inspection of Certain Optical Munitions of War', *Transactions of the Optical Society*, 20:4 (1919), pp. 97–100.
16. *OH*, vol. 1, part 1, p. 10.
17. Ibid., p. 53.
18. Ibid., pp. 53–8 supplies the contract procedures in the rest of this section.

19. TNA MUN 4/745, Orders Placed for Scientific and Optical Instruments &c., 1 August 1914 to 31 March 1917, illustrates the state of outstanding orders.

20. *OH*, vol. 11, part 4, 'Rifles', chs 2 and 3.

21. R. J. Q. Adams, *Arms and the Wizard; Lloyd George and the Ministry of Munitions 1915–1916* (London: Cassell, 1978), chs 1–3.

22. Trebilcock, 'War and the Failure of Industrial Mobilization: 1899 and 1914', p. 140.

23. Ibid.

24. *OH*, vol. 1, part 1, pp. 46–71 provides the basis for this section relating to the handling of contracts.

25. *OH*, vol. 1, part 1, p. 53.

26. Data extracted from prices charged in UGD 295/19/2/4 and UGD 295/19/2/5, Customer Order files 1914 and 1915.

27. TNA MUN 4/745, Orders placed for Scientific and Optical Instruments &c, 1 August 1914 to 31 March 1917, does not give any prices.

28. For examples of Barr & Stroud's efforts to maintain supplies, see UGD 295/4/113, H. D. Jackson to Chance Bros, 11 November 1914 and Jackson to LePersonne & Co., 9 December 1914; UGD 295/4/118, Jackson to Chance Bros, 18 June 1915; UGD 295/4/119, Jackson to Parra Mantois & Cie.,16 June 1915.

29. Adams, *Arms and the Wizard*, p. 18.

30. McBride, *Technological Change and the U.S. Navy*, p. 4.

31. Adams, *Arms and the Wizard*, p. 18 and then pp. 21–3.

32. Extracted from UGD 295/19/2/3, Customer Order files 1913, and UGD 295/11/1, audited accounts and balance sheet, 1913.

33. UGD 295/19/2/4, Customer Order files, sample French contracts, 1913 and War Office contracts, 1914.

34. UGD 295/19/2/4, and UGD 295/19/2/5, Customer Order files 1914 and 1915 provide the source material for the following section unless otherwise indicated.

35. Cited by MacLeod and MacLeod, 'Government and the Optical Industry in Britain', p. 171.

36. Moss and Russell, *Range and Vision*, pp. 65–7.

37. UGD 295/4/6, Letter book, Jackson to Director of Navy Contracts, 31 July 1914.

38. UGD 295/4/109, Letter book, a series of almost identical letters was sent out between 3 and 5 August 1914.

39. UGD 295/4/6, Letter book, Jackson to Secretary of the Admiralty, 6 August 1914.

40. See chapter 2 and 4 in this volume.

41. Science Museum Library (hereafter SML), Adam Hilger Collection: HILG 3/1, typescript 'Mr Twyman's lecture August 1944, p. 17.

42. SML HILG 3/1, typescript, p. 27.

43. UGD 295/4/110, Letter book, contains a series of letters from Jackson to Twyman which provides the material concerning Adam Hilger for the rest of this section unless otherwise indicated.

44. UGD 295/19/2/3, Customer Order file.

45. UGD 295/4/110, Letter book, Jackson to Twyman, 21 August 1914.

46. A. B. Dewar, *The Great Munition Feat 1914–1918* (London: Constable, 1921), p. 221.

47. H. Barty-King, *Eyes Right: The Story of Dollond & Aitchison Opticians 1750–1985* (London: Quiller Press, 1986), p. 125.

48. SML HILG 7/1, Twyman obituary reprint (unidentified), p. 270.

49. UGD 295/4/11, Letter book, Jackson to Beck, 26 October 1914; Jackson to Periscopic Prism Co., 6 October 1914; Jackson to Watson, 14 October 1914.

50. UGD 295/26/1/25, Barr to William Taylor, 3 June 1903 requesting help in obtaining optical workers and endorsing Taylor Hobson's small machine tools.

51. McConnell, *Instrument Makers to the World*, p. 73.

52. UGD 295/4/109, Letter book, Jackson to Taylor, Taylor & Hobson, 11 August 1914 and 18 August 1914.

53. UGD 295/4/110, Letter book, Jackson acknowledges advice from Taylor, Taylor & Hobson, 25 August 1914.

54. McConnell, *Instrument Makers to the World*, p. 75.

55. UGD 295/4/110, Letter book, Jackson to Twyman, various dates.

56. UGD 295/4/110 Jackson to Twyman, 30 August 1914.

57. UGD 295/4/744, Letter book, Jackson to Hilger, 9 August 1907.

58. UGD 295/4/112, Letter book, Jackson to Twyman, thanking him for his cooperation, 7 November 1914.

59. UGD 295/4/112, Jackson to Twyman, 9 December 1914.

60. UGD 295/4/112, Barr & Stroud to Standard Optical Co., 17 September 1914.

61. UGD 295/4/112 Barr & Stroud to Sir William Arrol & Co. (builders), 5 November 1914.

62. UGD 295/4/112, Letter book, Jackson to Twyman, 9 December 1914.

63. UGD 295/4/112, Letter book, Jackson to Twyman, 14 December 1914.

64. UGD 295/4/112, Letter book, Jackson to Twyman, 17 December 1914.

65. UGD 295/4/110, Letter book, Jackson to Edward Taylor, 10 December 1914, records novel methods.

66. UGD 295/4/114, Letter book, Jackson to Taylor, Taylor & Hobson, 8 January 1915.

67. Entries in the Letter books cease after this date, the last one being to Beck.

68. UGD 295/4/112, Letter book entries for early 1915.

69. UGD295/4/118, Letter book, Jackson to Taylor Hobson, 18 May 1915 confirming that Taylor Hobson is now the sole outside supplier of FT20 objective lenses, and 2 June 1915 asking for quotation for another 1,100 similar lenses.

70. From examination of progress details recorded in UGD 295/19/2/3 and UGD 295/19/2/4, Customer Order files 1914 and 1915, and from contract progress comments in PRO WO/745, Order and Supply List.

71. I. Skennerton, *The British Sniper* (Margate: Skennerton, 1984), pp. 41, 47.

72. Moss and Russell, *Range and Vision*, p. 73.

73. UGD 295, Unclassified material, Russell research notes, Barr & Stroud Letter book, vol. 99, and UGD 295/4/625, Letter book, Jackson to Ministry of Munitions, 20 July 1915.

74. UGD 295/4/11, Letter book, Jackson to Glasgow Tramways, 28 October 1914.

75. UGD 295/4/7, Letter book, Jackson to Director of Naval Construction, 7 August 1914.

76. Adams, *Arms and the Wizard*, ch. 7.

77. UGD 295, Unclassified material, Barr & Stroud Letter book, vol. 103, 16 December 1914.

78. UGD 295, Unclassified material, Russell research notes: this cites Barr & Stroud Letter book vol. 110, 31 October 1914.

79. *OH*, vol. 1, part 2, pp. 58, 59.

80. *OH*, vol. 1, part 2, p. 59.

81. UGD 295/4/110, Letter book, Jackson to French Military Attaché, 14 December 1914.

82. UGD 295/4/110, Letter book, Jackson to Secretary, War Office, 2 January 1915.

83. UGD 295/4/110, Letter book, Jackson to Secretary, War Office, 4 January 1915.
84. For deliveries to Greece, see UGD 295/19/2/3, Customer Order files 1913–15, and for prohibition of exports see PRO BT/55/23, Evidence to Engineering Industries Committee of Enquiry, evidence of Archibald Barr, 20 October 1916 and 16 November 1916.
85. For the difficulties of small-arms manufacture, see *OH*, vol. 11, part 4.

5 Industrial Mobilization: The Ministry of Munitions and its Relationship with the Industry

1. *OH*, 12 vols (London: HMSO, 1922), vol. 11, 'The Supply of Munitions', part 3, Optical Munitions and Glassware, p. 1.
2. MacLeod and MacLeod, 'Government and the Optical Industry in Britain', p. 165; Williams, *The Precision Makers*.
3. TNA, Records of the Board of Trade, BT 66/2/MMW11, Col. Wedgwood to Mr Booth CMG/5315, 20 August 1915, and *OH*, vol. 11, 'The Supply of Munitions', part 3, Optical Munitions and Glassware, p. 1.
4. *OH*, vol. 11, p. 1–7 provides source material for the rest of this section, unless otherwise indicated.
5. *Who Was Who 1929–1940* (London: A. and C. Black, 1941).
6. For secret patents, see T. H. O'Dell, *Inventions and Official Secrecy: A History of Secret Patents in the United Kingdom* (Oxford: Clarendon Press, 1994), especially chapters 4 and 5.
7. F. J. Cheshire, *The Modern Rangefinder* (London: Harrison & Sons, 1916).
8. Auerbach, *The Zeiss Works and the Carl Zeiss Stiftung in Jena*.
9. S. P. Thompson, 'Opto-Technics', *Journal of the Royal Society of Arts* (1902), pp. 518–27.
10. Personal communication from Dr Wolfgang Zimmer, Archive der Carl Zeiss Jena GmbH, 12 November 2004.
11. See MacLeod and MacLeod, 'Government and the Optical Industry in Britain', pp. 169, 170.
12. LCC Report.
13. *OH*, vol. 11, part 3, p. 1 supplies this and the following quotations.
14. TNA MUN 4/55, Control of optical firms 1915–16, Cheshire to Col. Wedgwood, War Office, 13 August 1915 provides the source material for the rest of this section.
15. Extracted from Anderson, Burnett and Gee, *Handlist of Scientific-Instrument Makers' Trade Catalogues*.
16. TNA MUN 4/55 Draft Report 19 October 1917.
17. The draft gives no indication of who were those 'in a position to judge'.
18. LCC Report, p. 11.
19. Great Britain, Government, Customs & Excise Department; *Annual Statement of the Trade of the United Kingdom with Foreign Countries and British Possessions* (HMSO, London, published annually). Hereafter *Annual Statement*.
20. These corresponded to the categories identified in the 1907 *Census of Production* (London: HMSO, 1907) except for spectacle lenses which the *Annual Statement* recorded under a separate heading.
21. Extracted from *Annual Statement*, 1911 to 1914.

22. D. Edgerton, *Science, Technology and the British Industrial 'Decline' 1870–1970* (Cambridge: Cambridge University Press, 1996), pp. 3–5 discusses the idea of relative decline and its significance.

23. Extracted from *Annual Statement*, 1911 to 1914, and PRO WO 395/3, Annual Reports of the Director of Army Contracts 1911 to 1914, and UGD 295/ 19/2/2 and UGD 295/19/2/3, Customer order files, 1911–14.

24. *OH*, vol. 11, part 3, p. 1.

25. See Thompson, 'Opto-Technics', pp. 518–27; and Anon., 'The Proposed Establishment of an Institute of Technical Optics', London: British Science Guild, 1914, pp. 31–4.

26. *Transactions of the Optical Society*, 20 (May 1919), obituary noting his death on 14 September 1918. He is absent from any edition of *Who's Who*.

27. TNA MUN 4/5006, Weekly Reports about Supply, Design and Production, box 1, June 1917 provides the source for the rest of this section, unless otherwise noted.

28. See chapter 5 in this volume for War Office policy on placing contracts and chapter 2 for experiences during the Boer War.

29. W. Rosenhain, 'Optical Glass', *Journal of the Royal Society of Arts*, 64 (4 August 1916), pp. 677–8.

30. TNA BT 66/1 (MMW 5) Chance Bros. to War Office, June 1915.

31. *OH*, vol. 11, part 3, p. 24.

32. Chance, *A History of the Firm of Chance Brothers*, p. 185.

33. TNA BT 66/2 (MMW11), CM 6 minute 8 July 1915.

34. Chance, *A History of the Firm of Chance Brothers*, p. 185.

35. Ibid., p. 183.

36. Ministry of Munitions, pp. 10, 11. Chance also had to pay the government a 5 per cent commission on all commercial sales.

37. Chance, *A History of the Firm of Chance Brothers*, p. 184.

38. TNA MUN 4/55, DDGC to DGMS, 27 July 1915.

39. TNA MUN 7/78, Instructions to Contractors, 13 May 1915, reminded suppliers of this obligation.

40. *OH*, vol. 11, part 3, p. 2.

41. TNA MUN 7/96, Wedgwood to DGMS, 23 March 1916. Cited by MacLeod and MacLeod, 'Government and the Optical Industry in Britain', p.176.

42. TNA MUN 7/78, Instructions to Contractors, 13 May 1915.

43. TNA BT 66/6/MMW47, 'Government Control of Industry: Report on the Manner in which Direct Control is Exercised'. 1918.

44. *OH*, vol. 3, part 3, p. 35.

45. UGD 295/4/111, Letter book, Jackson to Negretti & Zambra, 13 October 1914; UGD 295/4/112, Letter book, Jackson to Dollond, 27 November 1914.

46. *OH*, vol. 1, part 4, p. 1, provides the quotation; for further background information and detail on the Act, see *OH*, vol. 1, part 4, ch. 1; *OH*, vol. 3, part 3, ch. 2; Adams, *Arms and the Wizard*, chs 4, 5, 6 and 7.

47. *OH*, vol. 3, part 4, p. 17.

48. *OH*, vol. 3, part 3, p. 31 provides the source material for the rest of this section.

49. TNA MUN 4/55, Cheshire to Col. Wedgwood, 13 August 1915 is the source for this section.

50. Sources for the following are: for Barr & Stroud, UGD 295/11/1, Audited Accounts; for Thomas Cooke & Sons Ltd, University of York, Borthwick Institute, VIA, T. Cooke & Sons, AJB 030 1.1.1, Minute book 1897–1924; for Troughton & Sims Business

Records, VIA AJB 060 1.2.3, Balance sheets 1908–1919; for Taylor Hobson, Cooke Optics Ltd, Leicester, unclassified records, Taylor Hobson Directors' Minute Book No.1, entries dated 8 October 1912, 7 January 1913 and 11 March 1914.

51. McConnell, *Instrument Makers to the World*, ch. 5 describes Troughton & Sims' activities, but the author's suggestion of involvement in rangefinder production lacks evidence in support.

52. VIA AJB 020 1.2.3.

53. *OH*, vol. 3, part 3, p. 32; pages 31–5 provide the source material for the rest of this section on profits and taxation.

54. *OH*, vol. 3, part 3, p. 33.

55. Ibid., p. 34.

56. Ibid., p. 33.

57. See Williams (1994) pp. 63, 64, citing PRO MUN 4/55, Cheshire to Wedgwood, 13 August 1915.

58. TNA MUN 4/55, Control of optical firms, DDGC to DGMS, 27 July 1915.

59. TNA MUN 4/55, Control of optical firms, Wedgwood to Sir H. Llewellen Smith, 30 July 1915.

60. TNA MUN 4/55, Control of optical firms, supplies the source material for this preceding section: DDGC to DGMS, 27 July 1915; Beveridge to Sir H. Llewellen Smith, 30 July 1915; Wedgwood to Eric Geddes, 14 August 1915; Geddes to DGMS, 14 August 1915.

61. TNA MUN 4/745, Orders placed for Scientific and Optical Instruments &c, 1 August 1914 to 31 March 1917.

62. Hagen, 'Export versus Direct Investment in the German Optical Industry', p. 7.

63. TNA MUN 4/55, Control of optical firms provides the source material for this section: Memorandum of Meeting 17 January 1916, O. H. Smith to Esslemont 1 February1916, Wedgwood to Smith 4 February1916.

64. TNA MUN 4/55, Control of optical firms, Smith to Wedgwood 11 February 1916, Wedgwood to Smith 17. February 1916, Smith to DGMS 1 March 1916.

65. MacLeod and MacLeod, 'Government and the Optical Industry in Britain', p. 185.

66. TNA MUN 2/1a, Secret Weekly Reports, vol. 1, 18 September 1915, 23 October 1915 and 6 November 1915.

67. MacLeod and MacLeod, 'Government and the Optical Industry in Britain', p. 186.

68. *OH*, vol. 11, part 3, p. 109.

69. MacLeod and MacLeod, 'Government and the Optical Industry in Britain', p. 109.

70. TNA MUN 4/745, shows the extent to which existing patterns made up the bulk of orders.

71. See chapter 7 in this volume for details of the Aldis Brothers.

72. MacLeod and MacLeod, 'Government and the Optical Industry in Britain', p. 201.

73. Ibid., pp. 186, 201.

74. Leeds Industrial Museum, Armley Mills; Kershaw papers (unclassified), typescript by N. Kershaw, 'The History of Kershaws', p. 6 gives details of his prior lack of experience and training and the benefits of the course.

75. This is detailed in chapter 6 of this volume.

76. TNA MUN 4/5006, Reports of Technical Branch; Inspection of Labour, report 7 March 1918.

77. TNA MUN 4/5006, Reports of Technical Branch; Inspection of Labour, report 28, February 1918.

78. UGD 295/19/2/4, Customer Order files, CO 2965, 23 August 1916.
79. UGD 295/26/2/49, Bryson to Esslemont, 4 April 1916.
80. TNA MUN 4/5004, Weekly Reports, 8 January 1917.
81. UGD 295/4/634, Letter book, Barr & Stroud to Director of Inspection of Optical Supplies, 8 October 1918, 17 October 1918 and 8 November 1918 provide the source material for this section.
82. UGD 295/4/634, Letter book, Barr & Stroud to Controller of Optical Munitions Supply, 23 July 1918, 24 July 1918 and 27 July 1918 provides the source material for the rest of this section.
83. TNA MUN 4/5004 box 1 and MUN 4/5006, Weekly and other reports, provide the source material for this section, dates as given in the text.
84. TNA MUN 4/5006, Weekly Report, 16 August 1917.

6 The Industry's Wartime, 1915–18

1. MacLeod and MacLeod, 'Government and the Optical Industry in Britain', p. 166.
2. See T. P. Hughes, 'The Evolution of Large Technological Systems', in W. E. Bijker, T. P. Hughes and T. J. Pinch (eds), *The Social Construction of Technological Systems: New Directions in the Sociology and History of Technology* (Cambridge, MA: MIT Press, 1989), pp. 51–82, on pp. 73–5.
3. *OH*, vol. 11, 'The Supply of Munitions', part 3, p. 133, appendix 3(a).
4. See chapter 4 in this volume for the details of British firms. Also TNA, Records of the Ministry of Munitions, MUN 4/745 for samples of French contracts and *OH*, vol. 11, part 3, p. 133, appendix 3(a) and *OH*, vol. 1, part 1, appendix 2 for details of total orders placed.
5. *OH*, vol. 11, part 3, pp. 42–3.
6. In the absence of factory records, these figures are based on serial numbers taken from surviving instruments whose dates of manufacture can positively be placed in this period through reference to advertisements and the maker's own catalogues. I am particularly grateful to William Reid for providing data from his own records and collection.
7. See online at http://www.binoculars-cinecollectors.com/C.P._Goerz_Optics.pdf [accessed 16 January 2013].
8. These figures are provided partly from data published in Seeger, *Feldstecher*, pp. 102–4, and from further personal communications from Dr Seeger, as well as information provided by Thomas Antoniades.
9. Seeger, *Feldstecher*, ch. 4 and F. Watson, *Binoculars, Opera Glasses and Field Glasses* (Princes Risborough: Shire Books, 1995), pp. 13–19.
10. Auerbach, *The Zeiss Works and the Carl Zeiss Stiftung in Jena* describes the factory and its methods in detail.
11. MacLeod and MacLeod, 'Government and the Optical Industry in Britain', pp. 169, 170.
12. Hagen, 'Export versus Direct Investment in the German Optical Industry'.
13. Dewar, *The Great Munition Feat*, p. 217.
14. F. A. Carson, *Basic Optics and Optical Instruments* (Mineola, NY: Dover, 1997), pp. 10–32 and 10–45 explains the intricacies of binocular manufacture and the need for precise collimation.
15. See TNA MUN 4/745, Orders placed for Scientific and Optical Instruments and TNA MUN 4/5305 to TNA MUN 4/5313, Contract Cancellation files, for details.
16. Dewar, *The Great Munitions Feat*, p. 221 and TNA MUN 4/745, section on Ross Ltd, p. 45.

17. TNA MUN 4/745, section on W. Watson & Sons Ltd, p. 57.
18. A. W. Smith, *(Optical Works) Ltd 1850–1971: A Short History* (unpublished manuscript, n.d.), Bromley Local Archives Collection ref. L37.8/BN 107426), pp. 1–5 provides the source material for this section unless otherwise indicated.
19. TNA MUN 4/745, section on Aitchison & Co. Ltd, p. 2.
20. TNA MUN 4/5006, Reports on Optical Munitions Output, Technical Inspection and Labour Branch, Weekly Reports, 1915–1918: Report, 12 July 1917.
21. See Hagen, 'Export versus Direct Investment in the German Optical Industry', for background material to this.
22. Auerbach, *The Zeiss Works and the Carl Zeiss Stiftung in Jena*, p. 266.
23. Seeger, *Feldstecher*, pp. 83–100.
24. *OH*, vol. 11, part 3, p. 42. No source is given for this in the account, nor at the time of writing has any other evidence to support it yet been found.
25. H. T. Seeger, *Militarische Fernglaser Und Fernrohre* (Hamburg: Seeger, 1996), pp. 19–26 and pp. 83–100.
26. Williams, *The Precision Makers*, pp. 72–9.
27. TNA MUN 4/745, section on Bausch & Lomb, p. 7 and TNA MUN 4/5528, letter from A. S. Esslemont to Sherwood, 6 October 1916. See chapter 1 in this volume for comments on the influence of technological paradigms in the British Army. Many of the American binoculars were eventually offered to the Russians and never issued to British forces.
28. TNA MUN 4/672, Agreement between the Ministry of Munitions and Kershaw, 1916.
29. N. Channing and M. Dunn, *British Camera Makers; an A–Z Guide to Companies and Products* (Esher: Parkland Designs, 1996), p. 63.
30. Leeds Industrial Museum Library, Armley Mills, Leeds, W. Yorks, Kershaw material (unclassified): the papers left by Norman Kershaw (hereafter LIM/NK) provide the source material for the rest of this section unless otherwise indicated.
31. TNA MUN 4/745, section on Kershaw, p. 32.
32. Channing and Dunn, *British Camera Makers*, p. 63.
33. Adams, *Arms and the Wizard*, ch. 4, 'The men of push and go'.
34. *OH*, vol. 11, part 3, p. 18.
35. LIM/NK, p. 2.
36. *OH*, vol. 11, part 3, p. 26 provides source material for the rest of this section unless otherwise indicated.
37. TNA MUN 4/672, Agreement between the Ministry of Munitions and Kershaw. Unfortunately, only a summary of the actual document survives.
38. TNA MUN 5/312, Orders for Scientific Instruments &c., Binoculars, Prismatic, p. 5.
39. TNA MUN 4/745, for examples see sections on Beck, Ross and Watson.
40. *OH*, vol. 11, part 3, pp. 26–8 provides details.
41. MUN 4/5006, Reports of the Experimental Section, CM6, on optical munitions, various dates in 1917 and 1918.
42. *OH*, vol. 11, part 3, pp. 25–8, and MacLeod and MacLeod, 'Government and the Optical Industry in Britain', pp. 172–5.
43. TNA MUN 2/1, Secret Weekly Report, 29 March 1917.
44. TNA MUN 4/5006, Weekly Reports, 24 May 1917.
45. LIM/NK, p. 3, and Hagen, 'Export versus Direct Investment in the German Optical Industry', p. 7.
46. LIM/NK, p. 3.
47. LIM/NK, p. 3.

48. Dewar, *The Great Munition Feat*, p. 222.
49. TNA MUN 4/5006, Weekly Report 29 October 1918.
50. TNA MUN 4/745 provides source material for the rest of this section, unless otherwise indicated.
51. *OH*, vol. 11, part 3, p. 26.
52. TNA MUN 4/5006, Weekly Reports, 1918.
53. UGD 295/4/132, Letter book, H. D. Jackson to Brimfield, 15 March 1919.
54. See M. Pegler, *The Military Sniper since 1914* (Oxford: Osprey Publishing, 2001) for background information on the employment of snipers.
55. *OH*, vol. 11, part 3, p. 9 and MacLeod and MacLeod, 'Government and the Optical Industry in Britain', pp. 184–5.
56. See Skennerton, *The British Sniper*, chs 2 and 3 for background material. Skennerton's work drew heavily on Ministry of Munitions archive material.
57. Ibid., p. 34 illustrates examples, but leaves their manufacturers uncertain.
58. UGD 295/4/112, Letter book, Jackson to Chance Brothers, 7 December 1914, criticizes Aldis and UGD 295/4/110, Letter book, Jackson to Periscopic Prism Co., 25 August 1914, threatens to cancel all orders.
59. J. S. Carter, 'An Historical Analysis of the Development and Application of Visual and Aural Aids in English Education from 1900 to 1970' (PhD Thesis, University of Leeds, 1995), p. 292.
60. *Aldis Brothers & Their Productions* (Birmingham: Aldis Brothers, c. 1920.), p. 3.
61. *British Journal Photographic Almanac* (Liverpool: Henry Greenwood & Co., 1914), Aldis advertisement.
62. *Aldis Brothers*, p. 5.
63. Skennerton, *The British Sniper*, p. 47.
64. Ibid.
65. PRO MUN 4/745, section on Aldis Bros, p. 2.
66. TNA MUN 4/745, section on Periscopic Prism. Co., p. 41 provides contract details, and *OH*, vol. 11, part 3, p. 23 describes the take-over.
67. Pegler, *The Military Sniper since 1914*, p. 22.
68. Moss and Russell, *Range and Vision*, ch. 3.
69. These figures have been extracted from Customer Order records in UGD 295/19/2/4, UGD 295/19/2/5 and UGD 295/19/2/6.
70. Moss and Russell, *Range and Vision*, p. 80, *OH*, vol. 1, part 1, appendix 2, p. 145 and *OH*, vol. 11, part 3, appendix 3, p. 133.
71. From UGD 295/19/2/4, UGD 295/19/2/5 and UGD 295/19/2/6.
72. UGD 295/4/625, Letter book, Jackson to Esslemont, 27 December 1915.
73. UGD 295/26/1/14, Strang papers, historical notes concerning women workers 1916–1918, provide the source material for Barr & Stroud in the remainder of this section, unless otherwise indicated.
74. A. Marwick, *Women at War 1914–1918* (London: Fontana, 1977), p. 60.
75. Ibid.
76. Ibid., pp. 61–7.
77. UGD 295/19/2/6, Customer Order files, CO4000 of 22 August 1918 records an order for 300 artillery rangefinders for the US Army.
78. UGD 295/4/114, Letter book, Jackson to Keuffel & Esser, New York, 8 January 1915.

79. TNA BT 66/1/MMW 11, Nature of Demand for Optical Glass: minute from OMGD to Ministry of Munitions, 8 July 1915, warned of a likely embargo on French deliveries if rangefinder deliveries were again suspended.

80. UGD 295/4/10, Letter book, Jackson to Major Benson, Royal Artillery, Woolwich, 28 June 1915.

81. Extracted from UGD 295/19/2/3, UGD 295/19/2/3 and UGD 295/19/2/4, Customer Order files.

82. UGD 295/4/625, Letter book, Jackson to Director General Munitions Supply, 17 July 1915, acknowledging instructions.

83. UGD 295/4/625, Letter book, Jackson to Cheshire, 23 July 1915, acknowledging arrangements for visit.

84. UGD 295/4/119, Letter book, Jackson to Senechal, 12 July 1915, complaining of effects of War Office control after December 1915.

85. UGD 295/4/625, Letter book, Jackson to Esslemont, 9 August 1915.

86. UGD 295/4/625, Jackson to Esslemont, 18 September 1915.

87. UGD 295/4/625, Jackson to Esslemont, 21 September 1915.

88. For the preceding quotations see UGD 295/4/625, Jackson to Esslemont, 9 August 1915. For the learning curve, see UGD 295/4/625, Jackson to Esslemont, 22 October 1915; Jackson to Esslemont and Cheshire, 26 October 1915.

89. UGD 295/4/118, Letter book, C. Beck to Barr, 3 June 1915.

90. UGD 295/4/740, Letter book, Barr to C. Beck, 8 May 1916.

91. UGD 295/4/625, Letter book, Jackson to Secretary of Ministry of Munitions, return of workers employed, 20 July 1915.

7 Industrial Demobilization and Implosion, 1919

1. See L. S. Jaffe, *The Decision to Disarm Germany* (London: Allen & Unwin, 1985), chs 2 and 4.

2. See *OH*, vol. 3, part 2, pp. 112–43 for details of contract termination procedures.

3. TNA MUN 4/5308, Contract Liquidation Records, shows responses to cancellation notices dating from 13 November.

4. *OH*, vol. 2, part 1, 'Administrative Policy and Organisation', supplement, p. 16; pp. 15–19 provide source material for the following section, unless otherwise indicated, and pp. 20–42 other background material.

5. Dewar, *The Great Munition Feat*, p. 222 and TNA MUN 4/5308, letter R. & J. Beck to OMGD, 13 November 1918.

6. TNA MUN 4/5308, letter, Ross Optical Co. to OMGD, 19 November 1918.

7. TNA MUN 4/5308, letter, Hilger Ltd to OMGD, 21 November 1918.

8. Nothing seen in the PRO MUN contract cancellation files has shown any collaboration over cancellations prior to the November 1918 armistice.

9. TNA MUN 4/5308, minute, P. G. Henriques to H. A. Colefax, OMGD, 15 November 1918.

10. TNA MUN 4/5308, minute, G. Garnsey to H. A. Colefax, OMGD, 18 November 1918.

11. TNA MUN 4/5308, letter R. & J. Beck to Controller Optical Munitions Supply, 18 November 1918.

12. *OH*, vol. 1, section 2 details these safeguards.

13. TNA MUN 4/5313, Contract liquidations, summary of decisions made by Optical Munitions Liquidation Committee, 29 November 1918.

14. TNA MUN 4/5313, summary of decisions.

15. TNA MUN 4/5308, contract liquidations, sundry correspondence and MUN 4/5313, Liquidation Committee minute, 29 November 1918.

16. TNA MUN 4/5308, Controller OM to Sir W. Graham Greene, 4 December 1918. For later value see online at http://www.measuringworth.com/ukcompare/ [accessed 15 January 2013].

17. TNA MUN 4/745, Orders Placed for Scientific and Optical instruments etc. provides examples of such contracts and their delivery terms.

18. TNA MUN 4/5308, Liquidator's note to Liquidation Committee, 20 December 1918.

19. See chapters 6 and 7 in this volume.

20. *OH*, vol. 2, part 1, supplement, p. 17.

21. TNA MUN 4/5308, Edmund Batty to T. Knowles, OMGD, 10 January 1919 supplies this and the following quotation.

22. TNA MUN 4/5308 Minute, Controller to T. Knowles, 16 December 1918.

23. TNA MUN 4/5308 and 4/5309, Contract Liquidation papers provides the source material for this section unless otherwise indicated.

24. See *OH*, vol. 1, part 2 for background details of this aspect of the control of industry.

25. TNA MUN 4/5305 to /5313, Contract liquidation records contain much information. These figures are taken from TNA MUN 4/5308.

26. UGD, records of Barr & Stroud Ltd, 295/19/6/3, London Office papers, Admiralty Contract Cancellation file.

27. UGD 295/4/635, Letter book, H. D. Jackson to Controller of Optical Munitions Supply, 20 November 1918, provides material for the rest of this paragraph.

28. UGD 295/4/617, Letter book, Jackson to OMGD, 19 November 1918.

29. See chapter 7 in this volume for details about female workers.

30. UGD 295/16/1/58, Dr W. Strang's personal papers: works notices.

31. UGD 295/16/1/58, Dr W. Strang's personal papers: works notices.

32. Extracted from UGD 295/19/2/5 and UGD 295/19/2/5/6, Customer Order files 1916–19.

33. UGD 295/19/2/5 and 295/19/2/6, Customer Order files 1918–19.

34. Estimated from figures extracted from UGD 295/19/2/6 and UGD 295/19/2/7, Customer Order files.

35. Extracted from UGD 295/19/2/6, Customer Order files 1916–19.

36. UGD 295/4/634, Letter book, Jackson to R. T. Lacey at Woolwich, 19 November 1918.

37. See chapters 6 and 7 in this volume.

38. UGD 295/4/634, Letter book, Jackson to Director Munitions Accounts, 21 January 1919, provides the source material for the rest of this paragraph.

39. UGD 295/4/634. Letter book, Jackson to Liquidator of Optical Contracts, 27 January 1919 and UGD 295/4/634, Letter book, Jackson to Liquidator, Optical Munitions and Glassware Supply, 31 January 1919.

40. UGD 295/4/634, Letter book, Jackson to Liquidator, Optical Munitions and Glassware Supply, 4 March 1919.

41. UGD 295/19/8/3, War Office file, papers on liquidation of rangefinder contracts 1919, Jackson to Liquidator, 8 April 1919 and 29 April 1919.

42. Personal communication from Mr David Carson, head of Barr & Stroud's periscope department at the time of the move.

43. UGD 295/4/634, Letter book, Jackson to Liquidator, Optical Munitions and Glassware Supply, 29 March 1919.
44. Emphasis present in the original letter.
45. UGD 295/4/634, Letter book, Jackson to Liquidator, Optical Munitions and Glassware Supply, 2 May 1919.
46. UGD 295/19/6/3, Admiralty file, papers on liquidated and cancelled contracts, provides the source material for the following section, unless otherwise indicated.
47. UGD 295/19/6/3, Jackson to Director of Contracts, 8 August 1919.
48. UGD 295/19/6/3, Letter book, Admiralty to Barr & Stroud, 12 August 1919.
49. UGD 295/19/6/3, Letter book, Jackson to Admiralty, 18 August 1919.
50. UGD 295/11/1, Balance sheets, profit and loss accounts, Jackson's working papers.
51. Extracted from UGD 295/19/2/6 and UGD 295/19/2/7, Customer Order files 1919.
52. UGD 295/4/132, Letter book, J. W. French to A. Barr, 3 June 1919.
53. Moss and Russell, *Range and Vision*, ch. 4 and W. Reid, *We're Certainly not Afraid of Zeiss: Barr & Stroud Binoculars and the Royal Navy* (Edinburgh: National Museums of Scotland, 2001), ch. 1.
54. Reid, *We're Certainly not Afraid of Zeiss*, p. 22.
55. PRO WO 32/4947, minute, Secretary, War Cabinet to Secretary, War Office, 23 September 1919.
56. PRO WO 32/4947, minute, Secretary, War Cabinet to Secretary, War Office, 23 September 1919.
57. PRO MUN 4/165, MUN 4/166 and MUN 4/167, Weekly and Monthly Reports.
58. Reid, *We're Certainly not Afraid of Zeiss*, p. 22.
59. See Jaffe, *The Decision to Disarm Germany*, ch. 4.
60. Hagen, 'Export versus Direct Investment in the German Optical Industry'.
61. Seeger, *Feldstecher* and Watson, *Binoculars, Opera Glasses and Field Glasses* give background information.
62. See chapter 7 in this volume.
63. UGD 295/4/635, Letter book, Jackson to Controller of Optical Munitions Supply, 4 March 1919.
64. UGD 295 Unlisted material, J. M. Strang, manuscript, *The History of Barr & Stroud*, p. 124.
65. Moss and Russell, *Range and Vision*, p. 103.
66. Ibid., p. 104.
67. Leeds Industrial Museum Library, Kershaw material (unclassified), 'A Kershaw & Sons, Leeds: A brief history based on notes by Mr. Cecil Kershaw'.
68. UGD 295/19/2/7, Customer Order File 1919–20, 11 June 1919.
69. UGD 295/4/315, Letter book, Jackson to Barr, 9 September 1919.
70. Moss and Russell, *Range and Vision*, pp. 104–6.
71. Moss and Russel, *Range and Vision* brings out some aspects of this side of Barr's character in chapters 1 to 4.

8 Adaption and Survival, 1919–23

1. MacLeod and MacLeod, 'Government and the Optical Industry in Britain', Williams, *The Precision Makers*, chs 5 and 6 and McConnell, *Instrument Makers to the World*, pp. 76–80.
2. TNA WO 395/4, Annual Reports of the Director of Army Contracts, 1920–1 to 1923–4.

3. Harrison, *The Development of HM Submarines*, p 22.4–p. 22.11. Also, UGD, Records of Kelvin & Hughes Ltd, UGD 33/4/2, Inventories and valuations, 1919.

4. Glass, *Victorian Telescope Makers*, p. 213.

5. Williams, *The Precision Makers*, p. 106 and TNA MUN 4/5306, Contract cancellations and liquidation advances.

6. Glass, *Victorian Telescope Makers*, p. 214.

7. Ibid., Grubb to Innes, 27 March 1919.

8. PRO MUN 4/5306, Contract cancellations and liquidation advances, as well as chapter 7 in this volume.

9. Glass, *Victorian Telescope Makers*, p. 215.

10. Ibid., Grubb to Innes, 19 August 1921.

11. Ibid., p. 206.

12. H. C. King, *The History of the Telescope* (London: Charles Griffin & Co. Ltd, 1955), chs 15 and 17 provide background material. See also McConnell and Glass for comments on the economics of astro-telescope manufacture.

13. Cambridge University Library, Vickers Collection (hereafter CUL/VC): Document 739, Periscopes, correspondence with Sir Howard Grubb 1908–1912: Document 1003, Electric Boat Co. correspondence 1901–7.

14. Harrison, *The Development of HM Submarines*, p. 22.8.

15. Moss and Russell, *Range and Vision*, pp. 83–4.

16. UGD 295/19/2/5, Customer Order records 1917.

17. Little information about the company's capacity for optical work has been traced. No reference to optical manufacturing equipment is to be found in the firm's 1919 inventory. See UGD, Records of Kelvin & Hughes Ltd, UGD 33/4/2, Inventories and valuations, 1919.

18. Glass, *Victorian Telescope Makers*, p. 215, Grubb to Innes, 26 May 1921.

19. Ibid., p. 216, Innes to Secretary of State, Pretoria, 19 May 1923.

20. Ibid., p. 215, Frank Robbins report.

21. Ibid., p. 225.

22. See Cumbria Archives, Barrow-in-Furness: Vickers Material BDB 16/500, Handbooks and specifications, various dates from 1918 to 1927 detailing the periscopes fitted to various submarines.

23. J. D. Scott, *Vickers, A History* (London: Weidenfeld & Nicolson, 1962), p. 132.

24. Sumida, *In Defence of Naval Supremacy*, ch. 6 provides a full account of these events.

25. McConnell, *Instrument Makers to the World*, p. 74.

26. No mention of any such steps occurs in CUL/VC Document 1366, Directors' minute book, entries 1920 to 1924, and University of York, Borthwick Institute, VIA AJB 030/1.1.1, Cooke Directors' minute book, entries 1915 onwards.

27. McConnell, *Instrument Makers to the World*, p. 74.

28. Ibid., p. 77.

29. CUL/VC Document 1366, Directors' minute book, entry dated 30 September 1920.

30. McConnell, *Instrument Makers to the World*, p. 77.

31. VIA AJB 050/1.2.3,Troughton & Simms Balance Sheets 1914–19, Income Tax Papers and Stock Figures provide financial details, and E. Mennim, *Reid's Heirs: A Biography of James Simms Wilson* (Braunton: Merlin Books, 1990), p. 127 describes family relationships.

32. VIA AJB 070/1.3, Index to Optical Munitions Drawings, references 3524 and 3677.

33. VIA AJB 070/1.3, Index to Optical Munitions Drawings, references 3434 to 3492 and 3860.

34. See Barr & Stroud's previous experiences described in earlier chapters in this volume.
35. CUL/VC Document 1366, Directors' minute book, entries 30 September 1920, 24 February 1921, 19 July 1922, 28 September 1922, 23 January 1923, 28 March 1924 and 2 May 1924.
36. McConnell, *Instrument Makers to the World*, p. 79.
37. VIA AJB 070/1.3, Index to Optical Munitions Drawings, references 3905 and 3906.
38. UGD 295/4/336, Letter book, Harold Jackson to Vickers, 12 July1922.
39. UGD 295/4/334, Letter book, Harold Jackson to F. Morrison, 2 May1922.
40. Pollen, *The Great Gunnery Scandal*, chs 1–5 provide background material on Pollen and his association with Cooke's.
41. McConnell, *Instrument Makers to the World*, pp. 86–8.
42. *OH*, vol. 1, part 4, sections 1–5 describes these controls.
43. UGD 295/4/146, Letter book, Jackson to Secretary of the Admiralty, 6 December 1918.
44. Nothing in the contemporary letter books examined shows any knowledge of it.
45. Ministry of Defence, Naval Library, *Monthly Record of Principal Questions Dealt with by the Director of Naval Ordnance* (hereafter PQ), July to December 1918. Minute No. 192, 17 December 1918. Other minutes provide the source material for the rest of this section, unless otherwise indicated.
46. PQ, July to December 1918, Director of Naval Ordnance to Director of Navy Contracts, 21.12.1918.
47. UGD 295/4/146, Letter book, Jackson to Secretary of the Admiralty, regarding conference with Barr & Stroud, 26 February 1919.
48. UGD 295/4/146, Letter book, Jackson to Secretary of the Admiralty, 3 March 1919, is the source for this paragraph.
49. UGD 295/4/315 Letter book, Jackson to Satiolas, 10 September 1919.
50. Moss and Russell, *Range and Vision*, pp. 101, 102, 103.
51. UGD 295, Unclassified material, Strang material, financial papers, UGD 295/4/315, Letter book, Jackson to the Federation of British Industries, 11 October 1919 and UGD 295/4/131, Letter book, Jackson to Conrad Beck, 23 January 1919.
52. UGD 295/4/316, Letter book, Jackson to Francis Morrison, London office, summarizing recent board meeting.
53. UGD 295/4/148, Letter book, Jackson to Admiralty, 27 October 1919 and UGD 295/4/316, Letter book, Jackson to Coventry Ordnance Co. Ltd, 30 October 1919.
54. UGD 295/4/148, Letter book, Jackson to Admiralty, 13 November 1919, and UGD 295/4/148, Letter book, Jackson to Director of Admiralty Contracts, 25 November 1919.
55. UGD 295/4/316, Letter book, Jackson to W. H. Martin (Dutch agent), 2 December 1919.
56. UGD 295/11/1 Balance sheets, profit and loss accounts. Audited accounts for 1919 and later supply the data for this section.
57. Figure derived from http://www.measuringworth.com/ukcompare/ [accessed 15 January 2013].
58. TNA, ADM 212/46, Barr & Stroud correspondence.
59. UGD 295/11/1 Balance sheets, profit and loss accounts. Extracted from annual accounts 1913 to 1923, trading account section.
60. UGD 295/4/147, Letter book, Jackson to Admiralty, Advisor of Costs of Production, 20 October 1919.
61. UGD 295/4/147, Letter book, Jackson to Director of Naval Contracts, 18 August 1919.

62. UGD 295/4/145, Letter book, Jackson to Director of Naval Contracts, 29 October 1918 and UGD 295/4/148, Letter book, Jackson to Accountant General, 11 November 1919.

63. UGD 295/4/148: Jackson to Commander Ardill, DNO, 22 January 1920. Contractors involved in bankruptcy proceedings were automatically disbarred from the Admiralty's list of approved contractors.

64. UGD 295, Unclassified material, Russell research notes, Barr & Stroud Board Meeting, 19 February 1919.

65. TNA ADM 212/46, correspondence between Admiralty and Director of Scientific Research, December 1920, supplies the material for the rest of this section.

66. TNA ADM 212/46, undated memorandum.

67. TNA ADM 212/46, memorandum from Director of Scientific Research to Admiralty, December 1920.

68. TNA WO 395/4, Report of the Director of Army Contracts 1920–1921; 31 March 1921, p. 14.

69. TNA WO 395/4, Report of the Director of Army Contracts 1920–1921, p. 91.

70. TNA WO 395/4, Report of the Director of Army Contracts 1920–1921; 31 March 1921, p. 14.

71. TNA WO 395/4, Report of the Director of Army Contracts 1920–1921; 31 March 1921, p. 14.

72. UGD 295/4/149, Letter book, J. W. French to Capt. F. C. Dreyer, 15 September 1920.

73. UGD 295/19/2/7 and UGD 295/19/2/7, Customer Order files 1920 and 1921.

74. UGD 295/M11 Jackson to Barr, 16 February 1921.

75. UGD 295/M11 Jackson to Barr, 16 February 1921.

76. UGD 295/4/325, Letter book, Jackson to Barr, 16 February 1921.

77. UGD 295/4/325, Letter book, Jackson to Barr, 25 February 1921.

78. UGD 295/4/325, Letter book, Jackson to C. P. McCarthy, 16 February 1921, UGD 295/4/325, Letter book, 25 February 1921, Jackson to Tongue & Co., 2 February 1921, UGD 295/4/325, Letter book, 25 February 1921, Jackson to Barr.

79. UGD 295/4/325, Letter book, 16 February 1921, Jackson to C. P. McCarthy.

80. UGD 295/19/2/8 and UGD 295/19/2/9, Customer Order files 1919–22 provide data for the rest of this paragraph.

81. UGD 295/11/1, Balance sheets, profit and loss accounts, 1912–28: audited accounts and working papers.

82. A. Raven and J. Roberts, *British Battleships of World War Two: The Development and Technical History of the Royal Navy's Battleships and Battle Cruisers from 1911 to 1946* (London: Arms & Armour Press, 1976), p. 75.

83. Raven and Roberts, *British Battleships*, chs 1–5.

84. P. P. O'Brien, *British and American Naval Power: Politics and Policy 1900–1936* (Westport, CT: Praeger, 1998), chs 6 and 7 provide background material.

85. Raven and Roberts, *British Battleships*, p. 98.

86. UGD 295/4/315, Letter book, Jackson to Barr, 2 March 1921.

87. UGD 295/4/150, Letter book, J. W. French to Secretary of the Admiralty, 25 June 1921.

88. UGD 295/4/150, Letter book, J. W. French to Secretary of the Admiralty, 29 June 1921.

89. O'Brien, *British and American Naval Power*, p. 166.

90. Raven and Roberts, *British Battleships*, p. 108.

91. UGD 295/4/151, Letter book, Jackson to Commander F. Bennett, RN, 18 September 1921.

92. UGD 295/4/151, Letter book, Jackson to T. Y. Baker, 13 December 1921.

93. The data for this and the following section has been extracted from UGD 295/11/1, Balance sheets, profit and loss accounts and working papers, 1912–28, and UGD 295/19/2/8, 295/19/2/9 and 295/19/2/10, Customer Order files 1920–22.

94. W. Reid, 'Military Binoculars from Venlo', in G. Groenendijk (ed.), *A Farewell to Arms: Liber Amicorum in Honour of Jan Piet Puype, Former Senior Curator of the Army Museum Delft* (Delft: Legersmuseum, 2004), pp. 82, 83.

95. Bausch & Lomb Archives, Rochester New York, unclassified material (hereafter B&LA), G. S. Saegmuller, letters to Bausch & Lomb from Jena, Germany, various dates during 1920 supplies the source material for this paragraph.

96. UGD 295/4/637, Letter book, J. W. French to Col. Alison, Royal Artillery College, Woolwich, 4 July 1923.

97. Moss and Russell, *Range and Vision*, pp. 110, 111, and UGD 295/4/334, Letter book, Jackson to French, 31 May 1922.

98. B&LA, Saegmuller letters and J. Alexander, 'Nikon and the Sponsorship of Japan's Optical Industry by the Imperial Japanese Navy, 1917–1945', Department of History, University of British Columbia. Available online at http://grad.usask.ca/gateway/archive17.html [accessed 14 January 2013].

99. See chapter 4 in this volume.

100. UGD 295/4/154, Letter book, J. W. French to Commander Bruce Fraser, RN, 16 July 1923.

101. Moss and Russell, *Range and Vision*, p. 111 – more orders were received in 1924.

102. UGD 295/19/2/10, Customer Order files 1922–3.

WORKS CITED

Primary Sources

Manuscripts

Bausch & Lomb Inc., Rochester, New York, Company Archive, Saegmuller Letters (Unclassified material)

Bromley Library, Local Archives Collection, Bromley, Kent, Wray (Optical Works) Ltd, L37.8

Cambridge University Library, Vickers Collection, Document 739 (Grubb correspondence)

Cambridge University Library, Vickers Collection, Document 763 (Thomas Cooke & Sons Ltd)

Cambridge University Library, Vickers Collection, Document 1366 (Directors' minute book 8)

Cambridge University Library, Vickers Collection, Document 1367

Cumbria Archives, Barrow-in-Furness, Vickers Material, BDB 16 (Handbooks and specifications)

Hampshire County Record Office, Winchester, Priddy's Hard Material: Series, 109/M/91 (Admiralty gunnery branch)

HSBC Bank Group, Records of the Yorkshire Banking Company, Board Minutes 1882 to 1892

Leeds Industrial Museum Library, Armley Mills, Leeds, Kershaw Collection, Cecil Kershaw Papers (unclassified)

London Metropolitan Archives, London County Council, LCC/MIN (Education Committee minutes, 1911)

Ministry of Defence, Admiralty Library, Whitehall, London, *Monthly Record of Principal Questions dealt with by the Director of Naval Ordnance*

The National Archives, Kew, London, Admiralty Records, ADM 116 (Technical records)

The National Archives, Kew, London, Admiralty Records, ADM 212

The National Archives, Kew, London Board of Trade Records, BT 66 (Records of the Board of Trade)

The National Archives, Kew, London Ministry of Munitions Records, MUN 4 (Records of the Ministry of Munitions)

The National Archives, Kew, London Treasury Records, TI 11223 (Marindin case papers)

The National Archives, Kew, London War Office Records, WO 32

The National Archives, Kew, London War Office Records, WO 33 (Records of the Director of Artillery)

The National Archives, Kew, London War Office Records, WO 108 (Reports from South Africa, 1901)

The National Archives, Kew, London War Office Records, WO 395 (Records of the Director of Army Contracts)

Science Museum Library, South Kensington, London, Adam Hilger Archive, HILG 3/1 (Historical records)

University of Glasgow Archives, GB0248 Barr & Stroud Ltd Collection, UGD295, UGD 295/4 (Letter books)

University of Glasgow Archives, GB0248 Barr & Stroud Ltd Collection, UGD295, UGD 295/11 (Accounts and balance sheets)

University of Glasgow Archives, GB0248 Barr & Stroud Ltd Collection, UGD295, UGD 295/16 (Letters and personal papers)

University of Glasgow Archives, GB0248 Barr & Stroud Ltd Collection, UGD295, UGD 295/19 (Orders and contracts)

University of Glasgow Archives, GB0248 Barr & Stroud Ltd Collection, UGD295, UGD 295/22 (Patents and correspondence)

University of Glasgow Archives, GB0248 Barr & Stroud Ltd Collection, UGD295, UGD 295/26 (Personal papers)

University of Glasgow Archives, GB0248 Barr & Stroud Ltd Collection, UGD295, UGD 295, unclassified material (Russell & Strang notes)

University of Glasgow Archives, GB0248 Kelvin & Hughes Ltd Collection, UGD 33/4 (Inventories and valuations)

University of York, Borthwick Institute for Archives, Vickers Archive, Cooke, Troughton & Simms Ltd, AJB 070 (Company records)

University of York, Borthwick Institute for Archives, Vickers Archive, T. Cooke & Sons, AJB 030 (Minute book)

University of York, Borthwick Institute for Archives, Vickers Archive, T. Cooke & Sons, AJB 110 (Trade records)

University of York, Borthwick Institute for Archives, Vickers Archive, T. Cooke & Sons, AJB 210 (Miscellaneous papers)

University of York, Borthwick Institute for Archives, Vickers Archive, T. Cooke & Sons, AJB 220 (Papers written by employees)

University of York, Borthwick Institute for Archives, Vickers Archive, Troughton & Simms, AJB 050 (Company records)

Printed Sources

Census of Production 1907 (London: HMSO, 1907).

Great Britain, Admiralty, *Manual of Gunnery for H. M. Fleet, 1907* (London: HMSO, 1907).

Great Britain, Admiralty, *Manual of Gunnery for H. M. Fleet, 1915* (London: HMSO, 1915).

Great Britain, Admiralty, *Manual of Gunnery for H. M. Fleet, 1917* (London: HMSO, 1917).

Great Britain, Admiralty, *Rate Book for Naval Store: Authorised List and Price List of Naval Stores* (London: HMSO, annually from 1870).

Great Britain, Army, *Handbook of the Mekometer* (London: HMSO, 1911).

Great Britain, Army, *List of Changes in War Material 1891* (London: HMSO, 1891).

Great Britain, Army, *Regulations for Musketry Instruction, 1887* (London: HMSO, 1887).

Great Britain, Army, *Regulations for Musketry Instruction, 1896* (London: HMSO, 1896).

Great Britain, Army, School of Musketry, *Annual Report, 1891* (London: HMSO, 1891).

Great Britain, Army, School of Musketry, *Annual Report, 1893* (London: HMSO, 1893).

Great Britain, Government, Customs & Excise Department, *Annual Statement of the Trade of the United Kingdom with Foreign Countries and British Possessions* (London: HMSO, published annually).

Great Britain, Ministry of Munitions, *History of the Ministry of Munitions*, 12 vols (London: HMSO, 1922).

Secondary Sources

Aldis Brothers & Their Productions (Birmingham: Aldis Brothers, *c.* 1920).

Adams, R. J. Q., *Arms and the Wizard: Lloyd George and the Ministry of Munitions 1915–1916* (College Station, TX: A&M University Press and London: Cassell, 1978).

Alexander, J., 'Nikon and the Sponsorship of Japan's Optical Industry by the Imperial Japanese Navy, 1917–1945' (MA dissertation, University of British Columbia, 1999). Available online at http://grad.usask.ca/gateway/archive17.html [accessed 14 January 2013].

Anderson, R. G. W., J. Burnett and B. Gee, *Handlist of Scientific Instrument Makers' Trade Catalogues 1600–1914* (Edinburgh: National Museums of Scotland, 1990).

Anon., *The Century's Progress: Yorkshire Industry and Commerce 1893* (London: The London Printing and Engraving Co., 1893, reprint, Brenton Publishing, 1971).

Anon., 'The Proposed Establishment of an Institute of Technical Optics' (London: British Science Guild, 1914).

Auerbach, F., *The Zeiss Works and the Carl Zeiss Stiftung in Jena*, trans. F. Cheshire and S. Paul, 2nd edn (London: Marshall Brookes & Chalkely, 1904).

—, *The Zeiss Works*, trans. R. Kanthack (London: W. & G. Foyle, 1924).

Barr, A. and W. Stroud, *On Some New Telemeters, or Range-Finders* (London: British Association for the Advancement of Science, 1890).

Barty-King, H., *Eyes Right: The Story of Dollond & Aitchison Opticians 1750–1985* (London: Quiller Press, 1986).

Bastable, M. J., *Arms and the State: Sir William Armstrong and the Remaking of British Naval Power, 1854–1914* (Aldershot: Ashgate Publishing, 2004).

Bijker, W. E., T. P. Hughes and T. J. Pinc (eds), *The Social Construction of Technological Systems* (Cambridge, MA: MIT Press, 1989).

British Journal Photographic Almanac (Liverpool: Henry Greenwood & Co., 1914).

Brown, D. K., *The Grand Fleet: Warship Design and Development 1906–1922* (London: Chatham Publishing, 1999).

Callwell, C. E. and J. F. Headlam, *The History of the Royal Artillery*, 3 vols (London: Royal Artillery Institution, 1937).

Carson, F. A., *Basic Optics and Optical Instruments* (Mineola, NY: Dover, 1997).

Carter, J. S., 'An Historical Analysis of the Development and Application of Visual and Aural Aids in English Education from 1900 to 1970' (PhD thesis, University of Leeds, 1995).

Chance, J. F., *A History of the Firm of Chance Brothers & Co.* (London: Ballantyne & Co. Ltd, 1919).

Channing, N. and M. Dunn, *British Camera Makers; an A–Z Guide to Companies and Products* (Esher: Parkland Designs, 1996).

Cheetham, R. J., *Old Telescopes* (Southport, Lancashire: Samedie, 1997).

Cheshire, F. J., *The Modern Rangefinder* (London: Harrison & Sons, 1916).

Clark, T. N., A. D. Morrison-Low and A. D. C. Simpson, *Brass & Glass: Scientific Instrument Making Workshops in Scotland* (Edinburgh: National Museums of Scotland, 1989).

Dewar, A. B., *The Great Munition Feat 1914–1918* (London: Constable, 1921).

Edgerton, D., *Science, Technology and the British Industrial 'Decline' 1870–1970* (Cambridge: Cambridge University Press, 1996).

Edwards, E. G., 'Field Range-Finding', *Proceedings of the Royal Artillery Institution*, 11 (1883), pp. 202–14.

Forbes, G., *Experiences in South Africa with a New Infantry Range-finder* (London: Keliher & Co. Ltd, 1902).

Friedman, N., *U.S. Submarines through 1945: An Illustrated Design History* (Annapolis, MD: Naval Institute Press, 1995).

Garbett, H., *Naval Gunnery* (London: George Bell, 1897).

Glass, I. S., *Victorian Telescope Makers: The Lives and Letters of Thomas and Howard Grubb* (Bristol: Institute of Physics Publishing, 1998).

Gleichen, A, *The Theory of Modern Optical Instruments: A Reference Book for Physicists, Manufacturers of Optical Instruments and for Officers in the Army and Navy*, trans. H. Emsley and W. Swain (London: HMSO, 1918).

—, *The Theory of Modern Optical Instruments*, 2nd edn (London: HMSO, 1921).

Guild, British Science. Ninth Annual Report of the British Science Guild (London: British Science Guild, 1915).

Hagen, A., 'Export versus Direct Investment in the German Optical Industry', *Business History*, 4 (October 1996), pp. 1–20.

—, *Deutsche Direktinvestionen in Grossbritannien, 1871–1918* (Stuttgart: Steiner Verlag, 1997).

Harrison, A. N., *The Development of HM Submarines* (London: Ministry of Defence, 1979).

Hermann, D. G., *The Arming of Europe and the Making of the First World War* (Princeton, NJ: Princeton University Press, 1996).

Hogg, I. V. and L. F. Thurston, *British Artillery Weapons and Ammunition 1914–1918* (Shepperton: Ian Allen Ltd, 1972).

Horne, D. F., *Optical Instruments and their Applications* (Bristol: Adam Hilger Ltd, 1980).

Hughes, T. P., 'The Evolution of Large Technological Systems', in W. E. Bijker, T. P. Hughes and T. J. Pinch (eds), *The Social Construction of Technological Systems: New Directions in the Sociology and History of Technology* (Cambridge, MA: MIT Press, 1989), pp. 51–82.

Jaffe, L. S., *The Decision to Disarm Germany* (London: Allen & Unwin, 1985).

Johnson, B. K., 'The No. 7 Dial Sight, Mk. 2', *Transactions of the Optical Society*, 21:5 (1920), pp. 176–86.

King, H. C., *The History of the Telescope* (London: Charles Griffin & Co. Ltd, 1955).

Lambert, N. A., *Sir John Fisher's Naval Revolution* (Columbia, SC: University of South Carolina Press, 1999).

The Lens Collector's Vademecum, available online at http://antiquecameras.net/lensvademecum.html [accessed 28 January 2013].

Lloyd, E. W. and A. G. Hadcock, *Artillery: Its Progress and Present Position* (Portsmouth: J. Griffin & Co., 1893).

McBride, W., *Technological Change and the U.S. Navy 1865–1945* (Baltimore, MD: Johns Hopkins University Press, 2000).

McConnell, A., *Instrument Makers to the World: A History of Cooke, Troughton & Simms* (York: William Sessions Ltd, 1992).

MacLeod, R. and K. MacLeod, 'Government and the Optical Industry in Britain 1914–18', in J. M. Winter (ed.), *War and Economic Development* (Cambridge: Cambridge University Press, 1977), pp. 165–203.

Martin, L. C., *Optical Measuring Instruments: Their Construction, Theory and Use* (London: Blackie, 1924).

Marwick, A., *Women at War* (London: Fontana, 1977).

Mennim, E. J., *Reid's Heirs: A Biography of James Simms Wilson* (Braunton: Merlin Books, 1990).

Moss, M. and I. Russell, *Range and Vision: The First Hundred Years of Barr & Stroud* (Edinburgh: Mainstream Publishing, 1988).

Nolan, Captain, 'The Range-Finder', *Proceedings of the Royal Artillery Institution*, 2 (1874), pp. 161–207.

O'Brien, P. P., *British and American Naval Power: Politics and Policy 1900–1936* (Westport, CT: Praeger, 1998).

O'Dell, T. H., *Inventions and Official Secrecy: A History of Secret Patents in the United Kingdom* (Oxford: Clarendon Press, 1994).

Padfield, P., *Aim Straight: A Biography of Sir Percy Scott* (London: Hodder & Stoughton, 1966).

Pegler, M., *The Military Sniper since 1914* (Oxford: Osprey, 2001).

Pollen, A., *The Great Gunnery Scandal* (London: William Collins & Co. Ltd, 1980).

Raven, A. and J. Roberts, *British Battleships of World War Two: The Development and Technical History of the Royal Navy's Battleships and Battle Cruisers from 1911 to 1946* (London: Arms & Armour Press, 1976).

Reid, W., 'Binoculars in the Army, Part 1: 1856–1903', *Army Museum* (1981), pp. 10–23.

—, 'Binoculars in the Army, Part 1: 1904–1919', *Army Museum* (1982), pp. 15–30.

—, *We're Certainly not Afraid of Zeiss: Barr & Stroud Binoculars and the Royal Navy* (Edinburgh: National Museums of Scotland, 2001).

—, 'Military Binoculars from Venlo', in G. Groenendijk (ed.), *A Farewell to Arms: Liber Amicorum in Honour of Jan Piet Puype, Former Senior Curator of the Army Museum Delft* (Delft: Legersmuseum, 2004).

Rodgers, H. C. B., *Artillery through the Ages* (London: Seeley, Service, 1971).

Rosenhain, W., 'Optical Glass', *Journal of the Royal Society of Arts*, 64 (4 August 1916), pp. 677–8.

Sambrook, S. C., 'The British Armed Forces and their Acquisition of Optical Technology', in *Yearbook of European Administrative History* (Nomos Verlagsgesellschaft, Baden-Baden, 2008), pp. 139–64.

Scott, J. D., *Vickers, a History* (London: Weidenfeld & Nicolson, 1962).

Seeger, H T., *Feldstecher: Fernglaser Im Wandel Der Zeit* (Borken: Bresser-Optik, 1989).

—, *Militarische Fernglaser Und Fernrohre* (Hamburg: Seeger, 1996).

Skennerton, I., *The British Sniper* (Margate: Skennerton, 1984).

Smith, A. W., *Wray (Optical Works) Ltd 1850–1971: A Short History* (unpublished manuscript, n.d.).

Smith, G. and D. A. Atchison, *The Eye and Visual Optical Instruments* (Cambridge: Cambridge University Press, 1997).

Stanley, W. F., *Surveying and Levelling Instruments* (London: Spon, 1901).

Strachan, H., *The First World War: Volume 1: To Arms* (Oxford: Oxford University Press, 2001).

Stroud, W., *Early Reminiscences of the Barr and Stroud Rangefinders* (Glasgow: privately printed, c. 1936).

Suetter, M. F., *The Evolution of the Submarine Boat, Mine, and Torpedo* (Portsmouth: Griffin, 1908).

Sumida, J. T., *In Defence of Navel Supremacy: Finance, Technology, and British Naval Policy 1889–1914* (London: Routledge, 1993).

Taylor, E. W., 'The New Cooke–Pollen Rangefinder', *Journal of the United States Artillery*, 41:3 (May–June 1914), pp. 813–31.

Thompson, S. P., 'Opto-Technics', *Journal of the Royal Society of Arts* (1902), pp. 518–27.

Trebilcock, C., 'War and the Failure of Industrial Mobilisation', in J. M. Winter (ed.), *War and Economic Development: Essays in Memory of David Joslin* (Cambridge: Cambridge University Press, 1975), pp. 139–64.

—, *The Vickers Brothers: Armaments and Enterprise 1854–1914* (London: Europa Publishers, 1977).

Watson, F., *Binoculars, Opera Glasses and Field Glasses* (Princes Risborough: Shire Books, 1995).

Who Was Who 1929–1940 (London: A. and C. Black, 1941).

Williams, A. C., 'The Design and Inspection of Certain Optical Munitions of War', *Transactions of the Optical Society*, 20:4 (1919), pp. 97–120.

Williams, M. E. W., *The Precision Makers: A History of the Instruments Industry in Britain and France, 1870–1939* (London: Routledge, 1994).

Wilson, H. W., *Battleships in Action*, 2 vols (London: Conway Maritime Press, 1995).

Winter, J. M. (ed.), *War and Economic Development* (Cambridge: Cambridge University Press, 1975).

INDEX

For Product Safety Concerns and Information please contact our EU
representative GPSR@taylorandfrancis.com
Taylor & Francis Verlag GmbH, Kaufingerstraße 24, 80331 München, Germany

www.ingramcontent.com/pod-product-compliance
Ingram Content Group UK Ltd.
Pitfield, Milton Keynes, MK11 3LW, UK
UKHW021617240425

457818UK00018B/611

* 9 7 8 1 1 3 8 6 6 1 9 6 7 *